D0882662

Books by Randall Jarrell

POETRY

THE WOMAN AT THE WASHINGTON ZOO 1960

SELECTED POEMS 1955

THE SEVEN-LEAGUE CRUTCHES 1951

LOSSES 1948

LITTLE FRIEND, LITTLE FRIEND 1945

BLOOD FOR A STRANGER 1942

ESSAYS

A SAD HEART AT THE SUPERMARKET 1962

POETRY AND THE AGE 1953

FICTION

PICTURES FROM AN INSTITUTION 1954

A SAD HEART AT THE SUPERMARKET

Randall Jarrell

A SAD HEART AT THE SUPERMARKET

Essays & Fables

Atheneum : New York
1962

THESE ESSAYS, in a shorter or earlier form, have been printed in *The Saturday Evening Post, Figaro, Daedalus, The American Scholar, Art News, Mademoiselle* and *The New Republic;* and in *The Anchor Book of Stories, The Best Short Stories of Rudyard Kipling, Understanding Poetry and Wilderness of Ladies.* The author gratefully acknowledges permission to reprint.

Library of Congress Catalog card number 62–11681

Published simultaneously in Canada by
McClelland & Stewart Ltd.

Manufactured in the United States of America by
Kingsport Press, Inc., Kingsport, Tennessee

Designed by Harry Ford

First Edition

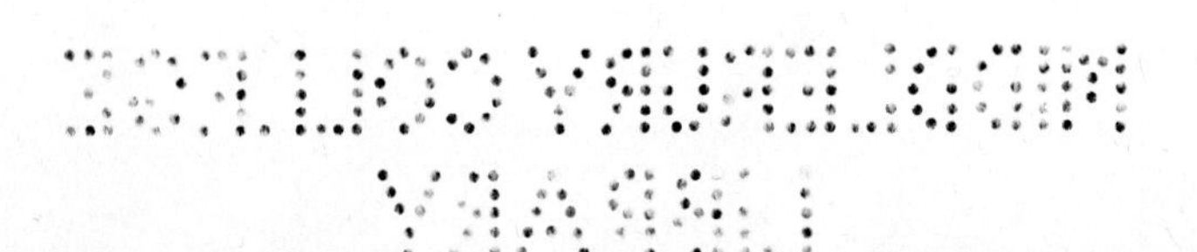

To Mary

The Author to the Reader

I've read that Luther said (it's come to me
So often that I've made it into meter):
And even if the world should end tomorrow
I still would plant my little apple-tree.
Here, reader, is my little apple-tree.

Contents

A SAD HEART AT THE SUPERMARKET

The Intellectual in America

THE philosopher Diogenes lived in a tub in the market place. He owned the clothes on his back and a wooden cup; one morning, when he saw a man drinking out of his hands, he threw away the cup. Alexander the Great came to Athens, and went down to the market place to see Diogenes; as he was about to leave he asked: "Is there anything I can do for you?" "Yes," said Diogenes, "you can get out of my light."

At different times, and in different places, this story has meant different things. The ages and places that have venerated wisdom, reason, lovers of wisdom—most ages and places have done so—have listened to the story and thought with wondering delight: "The things a man can do without!" Alexander may have owned Greece, Asia Minor, and part of Africa, but there was nothing he could do for Diogenes but move over and let the light fall on him. What is real in the world: that is what we must learn, Rilke wrote. Diogenes had learned; so that he could no longer be tipped or bribed with

Greece, Asia Minor, and part of Africa—with what the world thought reality, and he illusion. He had remained in his place, the place of wisdom, and had put Alexander the Great in his place, the place of power.

But when our age, our country, listens to the story of how Alexander stood in Diogenes' light, it asks perplexedly: "What was he doing *there?*" Why should a statesman, a general, make a sort of pilgrimage to a poverty-stricken philosopher, an intellectual of the most eccentric kind? We wouldn't. Most of us distrust intellectuals as such: we feel that they must be abnormal or else they wouldn't be intellectuals. This is so plain that a magazine like *Variety* can call our time "the era when to be accused of having some intellect is tantamount to vilification"; Brooks Atkinson, after noting that the American Psychological Association "has made the same point in more technical language," can conclude that "a passion for ignorance has swept the country." These passions never last; it is the settled marriage of convenience that troubles an American. A historian like Henry Steele Commager can say that "the historian of the future who chronicles this decade will be puzzled by the depth, strength, and prevalence of our anti-intellectualism," and can refer to "the vague aura of guilt that surrounds association with academic, intellectual, literary, and reform societies." When, a few years ago, men like McCarthy or Westbrook Pegler attacked or made fun of a man like Dean Acheson, they used as one of their most effective points against him the fact that he had—gone to Harvard.* Would their English or French or German counterparts have been able to use Oxford, the Sorbonne, or Heidelberg in the same way? Nor is it a question of party: plenty of Democrats would have done the same thing to a Republican Secretary of State; and when General Eisenhower defined an intellectual as "a man who takes more words than is necessary to tell more than he knows," he was speaking not as a Republican but as an American.

Haven't people got the story of Diogenes and Alexander backward? Didn't Diogenes wait, and wait, and wait? and, finally, go to Macedonia and get his Senator to make an appointment for him with the Emperor? and didn't the Palace Secretary say to the Senator, after seeing the week's schedule: Miss Macedonia, and the President of the Macedonian Federation of Labor, and the House Committee on Un-Macedonian Activities, and a delegation from the Macedonian Legion—didn't the Secretary say to Diogenes' Senator, as politely as he could: "The Emperor is a practical man, and has no time for philosophers"?

And then Diogenes went back to Athens. He had always been alone in his tub but, somehow,

* The reader will murmur with a smile: "That someone has gone to Harvard has rather a different point in 1962." Yes, doesn't it?—the same point that it had in 1932 or '42, under the second Roosevelt. Which of the points will it have in 1972 or '82? The tide goes in and the tide goes out, but the beach stays sand and the sea stays salt—and it is the sand and the salt that I am writing about.

he hadn't felt lonely: he had had for company the knowledge that someday Alexander would come—had had for company people's good will or good-humored indifference, their surprised or amused admiration, their resigned immemorial: "We may not have the sense or the time, but *someone* has to be wise." But now it was different. A Voice said to Diogenes, like the voice of God: "If some are wise, then others must be foolish: therefore I will have no one wise."

The Voice went on: "You highbrows, you long-hairs, you eggheads, are the way you are because there's something wrong with you. You sit there in your ivory tower"—but really it was a tub; where would Diogenes have got the money to buy a tower?—"pretending you're so different from other people, wasting your time on all these books nobody buys, and all these pictures my six-year-old boy can draw better than, and all these equations it takes another egghead like yourself to make heads or tails of—why don't you get wise to yourself and do what I do, and say what I say, and think like I think, and then maybe I'd have some respect for you?"

It was hard for Diogenes to know what to answer; and when he looked at his tub, it looked smaller and dingier than it used to look; and when he looked at the philosophy that had grown out of the tub, he felt about it the way an old Chinese poet said that he felt about his poetry: that if he wrote one of his poems on a cookie and gave it to a dog, the dog wouldn't eat it.

What could Diogenes do? Some people say he changed; changed until he was exactly the same as everybody else, only more so. Before long, people say, he owned the biggest advertising agency in Greece—or else it was the biggest broadcasting company. Or else both. People respected him, then. And every four years, late in the summer, Alexander the Great *would* come to see Diogenes; and as he was about to leave Diogenes would ask: "Is there anything I can do for you?" and Alexander would answer: "Well, yes. There're these speeches." Then Diogenes would write his speeches.

But some people say that Diogenes kept on the same as before, only he kept hearing voices—not voices exactly, but this Voice—and kept looking uneasily at people, as if they were about to do something to him or say something about him, when really they weren't paying any attention to him at all, except sometimes to laugh at him or to wonder to each other whether maybe he wasn't a Communist or else just crazy. There was a feeble-minded man in the market place that people used to laugh at and make jokes about, but people had got too civilized to make jokes about something like that any more, and they made them about Diogenes instead. And if you were a politician and something happened, you could blame it on Diogenes, part of the time—so he was useful to people, in a way; and sometimes Diogenes discovered things or invented things—penicillin, and television, and hybrid corn, and tensor analysis, and the hydrogen bomb—and wrote books and painted pictures and composed music and did all sorts of things that if you put a practical man in charge of, a business man, you could make a lot of money out of. The trouble with him wasn't that he was useless, exactly; it was more that he was—different.

One night Diogenes woke up and couldn't get back to sleep; he shifted back and forth in his tub, and repeated poems to himself, or said equations, or thought; finally he just lay there. And the Voice said to him, louder than he had ever heard it before:

"You are free to think differently from me and to retain your life, your property, and all that you possess; but you are henceforth a stranger among your people. You may retain your civil rights, but they will be useless to you, for you will never be chosen by your fellow citizens if you solicit their votes; and they will affect to scorn you if you ask for their esteem. You will remain among men, but you will be deprived of the rights of mankind. Your fellow creatures will shun you like an impure being; and even those who believe in your innocence will abandon you, lest they be shunned in their turn. Go in peace! I have given you your life, but it is an existence worse than death."

. . . But these last sentences were not said to Diogenes by some imaginary Voice, but were written a hundred and twenty-five years ago by Alexis de Tocqueville. This, he said, is what public opinion in the United States says to the man who disagrees with it. Many of this historian's statements about our country have a frightening and prophetic truth; and the passage of time has not altogether falsified the sentences which I have quoted. But things as they are in gross and confused reality are better than things as they were in Tocqueville's clear and penetrating imagination: he has created something which reality approaches as a limit. The American Diogenes is far better off inside Des Moines, or Jersey City, or Los Angeles, than inside Tocqueville's terrible sentences. He can become a celebrity, and be treated like other celebrities: we are willing to treat Hemingway and Faulkner as we treat Elvis Presley and Marilyn Monroe. And nowadays, after all, there are other people like Diogenes, some of whom say to him: *Brother;* there are people who, even if they are not themselves intellectuals, are willing for someone else to be; and—just as there are people who dislike Negroes or Jews or the Irish, but who like good Negroes, good Jews, good Irishmen, ones who are hardly like Negroes or Jews or Irishmen at all—there are people who dislike intellectuals but are willing to like a good intellectual, one who is hardly like an intellectual at all. And, too, there are the people of the rest of the world, most of whom tolerate, respect, admire even, intellectuals; it is a consolation to American intellectuals to know that their situation is, in some degree, a singular one.

They have suffered this misfortune: they are live, differing, individual human beings who have been put into a category that is itself a condemnation, of a kind—who are described sufficiently, people think, by an indicting stereotype. There is no way for them to free themselves of it. If we meet an honest and intelligent politician, a dozen, a hundred, we say that they aren't like politicians at all, and our category of politician stays unchanged: we know what politicians are. If a man thinks women men's intellectual inferiors, and keeps coming across women smarter than himself, he murmurs that the exception proves the rule, and saves for the first stupid woman he meets the scornful, categorizing: "Women!" We are this way about nationalities, faiths, races, sexes—about cats and dogs, even. And just as there are anti-Semitic Jews, women who despise women, there are intellectuals who enjoy attacking other intellectuals for being intellectuals. (Big fleas have little fleas to bite 'em, especially when the little ones know that they are going to get applauded by the dog.) And other intellectuals behave badly in other ways. It would be odd if they didn't. A looked-down-on class always gets some of its bad qualities simple from knowing that it is being looked down on; the calm and generosity and ease of the justly respected are replaced, often, by the uneasy resentment of the unjustly condemned. Toynbee says that the Turks took it for granted that the "Franks" among them possessed those qualities which the Franks, at home in Europe, considered ghetto qualities. If you have been put in your place long enough, you begin to act like the place. Some of the intellectual's faults are only our imagination, and some are our fault, and some are his fault. But his faults and his virtues, all his qualities, are more varied than we say. He is smart sometimes, stupid sometimes; ingenuous, disingenuous; nice, awful; so that we can say with perfect truth about this, as about so many things: "The more I see of intellectuals the less I know about the intellectual."

We are all—so to speak—intellectuals about something. General Eisenhower is an intellectual so far as military strategy is concerned: he has been taught, has taught himself, has read and thought and done, all the things that enable him to speak a language, think thoughts, make discriminations, that only other such intellectuals can fully understand or appreciate. If you want to be impressed with what an unintelligent amateur you are, with what trained, intelligent, and discriminating intellectuals the professionals are, sit in a hotel room with some coaches scouting a game, and hear what *they* have to say about the football game you thought you saw that afternoon. It takes a lot more than not being an intellectual to be right about anything. People are intellectuals about all kinds of things: if you know all about engines, why look with resentful distrust at someone who knows all about string quartets? Intellectuals are more like plain Americans than plain Americans think; plenty of them *are* plain Americans. And if they're complicated ones, different, is that really so bad? My daughter was telling me about a different boy, a queer one, whom all the other children looked down their noses at. I said, "How's he so different?" She said, "Lots of ways. He—he wears corduroys instead of blue jeans." Forgive us each day our corduroys.

Plain Americans enjoy telling Diogenes what they think of him; it would be interesting to know what he thinks of them. It is plain that, whether or not they like him, he likes them: he no longer despairs and flees to Europe, but stays home and suffers fairly willingly—is fairly thankful for—his native fate. Living among them as he does, he can hardly avoid realizing that Americans are a likable, even lovable people, possessing virtues some of which are rare in our time and some of which are rare in any time. But if he were to talk about the faults which accompany the virtues, he might say that the American, characteristically, thinks that nothing is hard or ought to be hard except business and sport; everything else must come of itself. Tocqueville said almost this, long ago: "His curiosity is at once insatiable and cheaply satisfied; for he cares more to know a great deal quickly than to know anything well. . . . The habit of inattention must be considered as the greatest defect of the democratic character." And he goes on to say that the American's leaders—whom Tocqueville calls, oddly, his courtiers and flatterers—"assure him that he possesses all the virtues without having acquired them, or without caring to acquire them." Diogenes could say to us: "You are not willing to labor to be wise—you are not even willing to be wise. It would be a change, and you are not willing to change: it would make you different from other Americans, and you are not willing to be different from them in any way. You wish to remain exactly as you are, and to have the rest of the world change until it is exactly like you; and it seems to you unreasonable, even perverse, for the rest of the world not to wish this too."

All this is very human. But it is very human, too, for the rest of the world, Europeans especially, to be afraid that we shall be successful in transforming them into what so many of them believe us to be: rich, powerful, and skillful barbarians, materialists who neglect or despise things of the mind and spirit. The American way of life, to many Europeans, means McCarthyism, comic books, Mickey Spillane, *et cetera, et cetera, et cetera.* If we say: "But they aren't the real America," and name the scientists and artists and scholars who seem to us the real America, these Europeans will answer: "They! Why, you look askance at them, attack or make fun of them—how gladly you would be rid of them!" Then we shall have to explain that they are taking a few remarks too seriously: that our country—the most advanced, technologically and industrially, that the world has ever known—has to depend for every moment of its existence upon the work of millions of highly educated specialists of every sort. We may not praise them, but we use them. Where should we be without the productions of intellectuals, the fruits of intellect?

Perhaps we need to let our allies know more about American culture, so that they can feel more as if they were accompanying a fellow and less as if they were following a robot. Perhaps we need to let more of our own people know about it and share it. Nobody ever before had so much money to spend, so much time to spend—do we spend it as interestingly and imaginatively as we might? Is what Tocqueville said so long ago, true today: that Americans "carry very low tastes into their extraordinary fortunes, and seem to have acquired the supreme power only to minister to their coarse and paltry pleasures"? Is it true that "the love of well-being now has become the predominant taste of the nation"? Do Americans, democratic peoples in general, need nothing so much as "a more enlarged idea of themselves and their kind"?

The Founding Fathers of our country were men who had an "enlarged idea of themselves and their kind"; they had for themselves and us great expectations. Franklin and Jefferson and Adams were men who respected, who labored to understand, and who made their own additions to, science and philosophy and education, the things of the mind and of the spirit. They would have disliked the word intellectual, as we may dislike it, because it seems to set apart from most men what it is natural and laudable for all men to aspire to—our species is called *homo sapiens;* but they would have admitted that, if you wanted to use the word, they were intellectuals. To look down upon, to stigmatize as eccentric or peripheral, science and art and philosophy, human thought, would have seemed to them un-American. It would not have seemed to them, even, human.

That most human and American of presidents—of Americans—Abraham Lincoln, said as a young man: "The things I want to know are in books; my best friend is the man who'll get me a book I ain't read." It's a hard heart, and a dull one, that doesn't go out to that sentence. The man who will make us see what we haven't seen, feel what we haven't felt, understand what we haven't understood—he *is* our best friend. And if he knows more than we do, that is an invitation to us, not an indictment of us. And it is not an indictment of him, either; it takes all sorts of people to make a world—to make, even, a United States of America.

The Taste of the Age

WHEN we look at the age in which we live —no matter what age it happens to be— it is hard for us not to be depressed by it. The taste of the age is, always, a bitter one. "What kind of a time is this when one must envy the dead and buried!" said Goethe about his age; yet Matthew Arnold would have traded his own time for Goethe's almost as willingly as he would have traded his own self for Goethe's. How often, after a long day witnessing elementary education, School Inspector Arnold came home, sank into what I hope was a Morris chair, looked round him at the Age of Victoria, that Indian Summer of the Western World, and gave way to a wistful, exacting, articulate despair!

Do people feel this way because our time is worse than Arnold's, and Arnold's than Goethe's, and so on back to Paradise? Or because forbidden fruits— the fruits forbidden us by time—are always the sweetest? Or because we can never compare our own age with an earlier age, but only with books about that age?

We say that somebody doesn't know what he is missing; Arnold, pretty plainly, didn't know what he was having. The people who live in a Golden Age usually go around complaining how yellow everything looks. Maybe we too are living in a Golden or, anyway, Gold-Plated Age, and the people of the future will look back at us and say ruefully: "We never had it so good." And yet the thought that they will say this isn't as reassuring as it might be. We can see that Goethe's and Arnold's ages weren't as bad as Goethe and Arnold thought them: after all, they produced Goethe and Arnold. In the same way, our times may not be as bad as we think them: after all, they have produced us. Yet this too is a thought that isn't as reassuring as it might be.

A Tale of Two Cities begins by saying that the times were, as always, "the best of times, the worst of times!" If we judge by wealth and power, our times are the best of times; if the times have made us willing to judge by wealth and power, they are the worst of times. But most of us still judge by more: by literature and the arts, science and philosophy, education. (Really we judge by more than these: by love and wisdom; but how are we to say whether our own age is wiser and more loving than another?) I wish to talk to you for a time about what is happening to the audience for the arts and literature, and to the education that prepares this audience, here in the United States.

In some ways this audience is improving, has improved, tremendously. Today it is as easy for us to get *Falstaff* or *Boris Godunov* or *Ariadne auf Naxos*, or Landowska playing *The Well-Tempered Clavichord*, or Fischer-Dieskau singing *Die Schöne Müllerin*, or Richter playing Beethoven's piano sonatas, as it used to be to get Mischa Elman playing *Humoresque*. Several hundred thousand Americans bought Toscanini's recording of Beethoven's Ninth Symphony. Some of them played it only to show how faithful their phonographs are; some of them played it only as the stimulus for an hour of random, homely rumination. But many of them really listened to the records—and, later, went to hear the artists who made the records—and, later, bought for themselves, got to know and love, compositions that a few years ago nobody but musicologists or musicians of the most advanced tastes had even read the scores of. That there are sadder things about the state of music here, I know; still, we are better off than we were twenty-five or thirty years ago. Better off, too, so far as the ballet is concerned: it is our good fortune to have had the greatest influence on American ballet the influence of the greatest choreographer who ever lived, that "Mozart of choreographers" George Balanchine.

Here today the visual arts are—but I don't know whether to borrow my simile from the Bible, and say *flourishing like the green bay tree*, or to borrow it from Shakespeare and say *growing like a weed.* We are producing paintings and reproductions of paintings, painters and reproductions of painters, teachers and museum-directors and gallery-goers and patrons of the arts, in almost celestial quantities. Most of the painters are bad or mediocre, of course —this is so, necessarily, in any art at any time— but the good ones find shelter in numbers, are bought, employed, and looked at like the rest. The people of the past rejected Cézanne, Monet, Renoir, the many great painters they did not understand; by liking and encouraging, without exception, all the painters they do not understand, the people of the present have made it impossible for this to happen again.

Our society, it turns out, can use modern art. A restaurant, today, will order a mural by Miro in as easy and matter-of-fact a spirit as, twenty-five years ago, it would have ordered one by Maxfield Parrish. The president of a paint factory goes home, sits down by his fireplace—it *looks* like a chromium acquarium set into the wall by a wall-safe company that has branched out into interior decorating, but there is a log burning in it, he calls it a fireplace, let's call it a fireplace too—the president sits down, folds his hands on his stomach, and stares relishingly at two paintings by Jackson Pollock that he has hung on the wall opposite him. He feels at home with them; in fact, as he looks at them he not only feels at home, he feels as if he were back at the paint factory. And his children—if he has any —his children cry for Calder. He uses thoroughly advanced, wholly non-representational artists to design murals, posters, institutional advertisements: if we have the patience (or are given the opportunity) to wait until the West has declined a little longer, we shall all see the advertisements of Merrill Lynch, Pierce, Fenner, and Smith illustrated by Jean Dubuffet.

This president's minor executives may not be willing to hang a Kandinsky in the house, but they will wear one, if you make it into a sport shirt or a pair of swimming-trunks; and if you make it into a sofa, they will lie on it. They and their wives and children will sit on a porcupine, if you first exhibit it at the Museum of Modern Art and say that it is a chair. In fact, there is nothing, nothing in the whole world that someone won't buy and sit in if you tell him that it is a chair: the great new art form of our age, the one that will take anything we put in it, is the chair. If Hieronymus Bosch, if Christian Morgenstern, if the Marquis de Sade were living at this hour, what chairs they would be designing!

Our architecture is flourishing too. Even colleges have stopped rebuilding the cathedrals of Europe on their campuses; and a mansion, today, is what it is not because a millionaire has dreamed of the Alhambra, but because an architect has dreamed of the marriage of Frank Lloyd Wright and a silo. We

Americans have the best factories anyone has ever designed; we have many schools, post-offices, and public buildings that are, so far as one can see, the best factories anyone has ever designed; we have many delightful, or efficient, or extraordinary houses. The public that lives in the houses our architects design—most houses, of course, are not designed, but just happen to a contractor—this public is a broad-minded, tolerant, adventurous public, one that has triumphed over inherited prejudice to an astonishing degree. You can put a spherical plastic gas-tower on aluminum stilts, divide it into rooms, and quite a few people will be willing to crawl along saying, "Is this the floor? Is this the wall?"—to make a down-payment, and to call it home. I myself welcome this spirit, a spirit worthy of Captain Nemo, of Rossum's Universal Robots, of the inhabitants of the Island of Laputa; when in a few years some young American airmen are living in a space-satellite part way to the moon, more than one of them will be able to look around and think: "It's a home just like Father used to make," if his father was an architect.

But in the rest of the arts, the arts that use words—

But here you may interrupt me, saying: "You've praised or characterized or made fun of the audience for music, dancing, painting, furniture, and architecture, yet each time you've talked only about the crust of the pie, about things that apply to hundreds of thousands, not to hundreds of millions. Most people don't listen to classical music at all, but to rock-and-roll or hillbilly songs or some album named Music To Listen To Music By; they've never seen any ballet except a television ballet or some musical comedy's last echo of *Rodeo*. When they go home they sit inside chairs like imitation-leather haystacks, chairs that were exhibited not at the Museum of Modern Art but at a convention of furniture dealers in High Point; if they buy a picture they buy it from the furniture dealer, and it was the furniture dealer who painted it; and their houses are split-level ranch-type rabbit-warrens. Now that you've come to the 'arts that use words,' are you going to keep on talking about the unhappy few, or will you talk for a change about the happy many?"

I'll talk about the happy many; about the hundreds of millions, not the hundreds of thousands. Where words and the hundreds of thousands are concerned, plenty of good things happen—though to those who love words and the arts that use them, it may all seem far from plenty. We do have good writers, perhaps more than we deserve—and good readers, perhaps fewer than the writers deserve. But when it comes to tens of millions of readers, hundreds of millions of hearers and viewers, we are talking about a new and strange situation; and to understand why this situation is what it is, we need to go back in time a little way, back to the days of Matthew Arnold and Queen Victoria.

II

We all remember that Queen Victoria, when she died in 1901, had never got to see a helicopter, a television set, penicillin, an electric refrigerator; yet she *had* seen railroads, electric lights, textile machinery, the telegraph—she came about midway in the industrial and technological revolution that has transformed our world. But there are a good many other things, of a rather different sort, that Queen Victoria never got to see, because she came at the very beginning of another sort of half-technological, half-cultural revolution. Let me give some examples.

If the young Queen Victoria had said to the Duke of Wellington: "Sir, the Bureau of Public Relations of Our army is in a deplorable state," he would have answered: "What is a Bureau of Public Relations, ma'am?" When he and his generals wanted to tell lies, they had to tell them themselves; there was no organized institution set up to do it for them. But of course Queen Victoria couldn't have made any such remark, since she too had never heard of public relations. She had never seen, or heard about, or dreamed of an advertising agency; she had never seen—unless you count Barnum—a press agent; she had never seen a photograph of a sex-slaying in a tabloid—had never seen a tabloid. People gossiped about her, but not in gossip columns; she had never heard a commentator, a soap opera, a quiz program. Queen Victoria—think of it!—had never heard a singing commercial, never seen an advertisement beginning: *Science says . . .* and if she *had* seen one she would only have retorted: "And what, pray, does the Archbishop of Canterbury say? What does dear good Albert say?"

When some comedian or wit—Sydney Smith, for example—told Queen Victoria jokes, they weren't supplied him by six well-paid gag-writers, but just occurred to him. When Disraeli and Gladstone made speeches for her government, the speeches weren't written for them by ghost-writers; when Disraeli and Gladstone sent her lovingly or respectfully inscribed copies of their new books, they had written the books themselves. There they were, with the resources of an empire at their command, and they wrote the books themselves! And Queen Victoria had to read the books herself: nobody was willing—or able—to digest them for her in *Readers' Digest*, or to make movies of them, or to make radio- or television-programs of them, so that she could experience them painlessly and effortlessly. In those days people chewed their own food or went hungry; we have changed all that.

Queen Victoria never went to the movies and had an epic costing eight million dollars injected into her veins—she never went to the movies. She never read a drugstore book by Mickey Spillane; even if she had had a moral breakdown and had read a Bad Book, it would just have been *Under Two Flags* or something by Marie Corelli. She had never been interviewed by, or read the findings of, a Gallup

Poll. She never read the report of a commission of sociologists subsidized by the Ford Foundation; she never Adjusted herself to her Group, or Shared the Experience of her Generation, or breathed a little deeper to feel herself a part of the Century of the Common Man—she *was* a part of it for almost two years, but she didn't know that that was what it was.

And all the other people in the world were just like Queen Victoria.

Isn't it plain that it is all *these* lacks that make Queen Victoria so old-fashioned, so finally and awfully different from us, rather than the fact that she never flew in an airplane, or took insulin, or had a hydrogen bomb dropped on her? Queen Victoria in a D. C. 7 would be Queen Victoria still—I can hear her saying to the stewardess: "We do not wish dramamine"; but a Queen Victoria who listened every day to *John's Other Wife*, *Portia Faces Life*, and *Just Plain Bill*—that wouldn't be Queen Victoria at all!

There has been not one revolution, an industrial and technological revolution, there have been two; and this second, cultural revolution might be called the Revolution of the Word. People have learned to process words too—words, and the thoughts and attitudes they embody: we manufacture entertainment and consolation as efficiently as we manufacture anything else. One sees in stores ordinary old-fashioned oatmeal or cocoa; and, side by side with it, another kind called Instant Cocoa, Instant Oats.

Most of our literature—I use the word in its broadest sense—is Instant Literature: the words are short, easy, instantly recognizable words, the thoughts are easy, familiar, instantly recognizable thoughts, the attitudes are familiar, already-agreed-upon, instantly acceptable attitudes. And if all this is true, can these productions be either truth or—in the narrower and higher sense—literature? The truth, as everybody knows, is sometimes complicated or hard to understand; is sometimes almost unrecognizably different from what we expected it to be; is sometimes difficult or, even, impossible to accept. But literature is necessarily mixed up with truth, isn't it?—our truth, truth as we know it; one can almost define literature as the union of a wish and a truth, or as a wish modified by a truth. But this Instant Literature is a wish reinforced by a cliché, a wish proved by a lie: Instant Literature—whether it is a soap opera, a Broadway play, or a historical, sexual best-seller—tells us always that life is not only what we wish it, but also what we think it. When people are treating him as a lunatic who has to be humored, Hamlet cries: "They fool me to the top of my bent"; and the makers of Instant Literature treat us exactly as advertisers treat the readers of advertisements—humor us, flatter our prejudices, pull our strings, show us that they know us for what they take us to be: impressionable, emotional, ignorant, somewhat weak-minded Common Men. They fool us to the top of our bent—and if we aren't fooled, they dismiss us as *a statistically negligible minority*.

An advertisement is a wish modified, if at all, by the Pure Food and Drug Act. Take a loaf of ordinary white bread that you buy at the grocery. As you eat it you know that you are eating it, and not the blotter, because the blotter isn't so bland; yet in the world of advertisements little boys ask their mother not to put any jam on their bread, it tastes so good without. This world of the advertisements is a literary world, of a kind: it is the world of Instant Literature. Think of some of the speeches we hear in political campaigns—aren't they too part of the world of Instant Literature? And the first story you read in the *Saturday Evening Post*, the first movie you go to at your neighborhood theater, the first dramatic program you hear on the radio, see on television—are these more like *Grimm's Tales* and *Alice in Wonderland* and *The Three Sisters* and *Oedipus Rex* and Shakespeare and the Bible, or are they more like political speeches and advertisements?

The greatest American industry—why has no one ever said so?—is the industry of using words. We pay tens of millions of people to spend their lives lying to us, or telling us the truth, or supplying us with a nourishing medicinal compound of the two. All of us are living in the middle of a dark wood—a bright Technicolored forest—of words, words, words. It is a forest in which the wind is never still: there isn't a tree in the forest that is not, for every moment of its life and our lives, persuading or ordering or seducing or overawing us into buying this, believing that, voting for the other.

And yet, the more words there are, the simpler the words get. The professional users of words process their product as if it were baby food and we babies: all we have to do is open our mouths and swallow. Most of our mental and moral food is quick-frozen, pre-digested, spoon-fed. E. M. Forster has said: "The only thing we learn from spoon-feeding is the shape of the spoon." Not only is this true—pretty soon, if anything doesn't have the shape of that spoon we won't swallow it, we can't swallow it. Our century has produced some great and much good literature, but the habitual readers of Instant Literature cannot read it; nor can they read the great and good literature of the past.

If Queen Victoria had got to read the *Readers' Digest*—awful thought!—she would have loved it; and it would have changed her. Everything in the world, in the *Readers' Digest*—I am using it as a convenient symbol for all that is like it—is a palatable, timely, ultimately reassuring anecdote, immediately comprehensible to everybody over, and to many under, the age of eight. Queen Victoria would notice that Albert kept quoting, from Shakespeare—that the Archbishop of Canterbury kept quoting, from the Bible—things that were very different from anything in the *Readers' Digest*. Sometimes these sentences were not reassuring but disquieting, sometimes they had big words or hard thoughts in them, sometimes the interest in them wasn't human, but literary or divine. After a while Queen Victoria would want Shakespeare and the Bible—would want Albert, even—digested for her beforehand by the *Readers' Digest.* And a little further on in this process of digestion, she would look from the *Readers' Digest* to some magazine the size of your palm, called *Quick* or *Pic* or *Click* or *The Week in TV*, and a strange half-sexual yearning would move like musk through her veins, and she would—

But I cannot, I will not say it. You and I know how she and Albert will end: sitting before the television set, staring into it, silent; and inside the set, there are Victoria and Albert, staring into the television camera, silent, and the master of ceremonies is saying to them: "No, I think you will find that *Bismarck* is the capital of North Dakota!"

But for so long as she still reads, Queen Victoria will be able to get the Bible and Shakespeare—though not, alas! Albert—in some specially prepared form. Fulton Oursler or Fulton J. Sheen or a thousand others are always re-writing the Bible; there are many comic-book versions of Shakespeare; and only the other day I read an account of an interesting project of re-writing Shakespeare "for students":

Philadelphia, Pa. Feb. 1. (AP)

Two high school teachers have published a simplified version of Shakespeare's "Julius Caesar" and plan to do the same for "Macbeth." Their goal is to make the plays more understandable to youth.

The teachers, Jack A. Waypen and Leroy S. Layton, say if the Bible can be revised and modernized why not Shakespeare? They made 1,122 changes in "Julius Caesar" from single words to entire passages. They modernized obsolete words and expressions and substituted "you" for "thee" and "thou."

Shakespeare had Brutus say in Act III, Scene I:

> *Fates, we will know your pleasures;*
> *That we shall die, we know; 'tis but the time*
> *And drawing days out, that men stand upon.*

In the Waypen-Layton version, Brutus says:

> *We will soon know what Fate decrees for us.*
> *That we shall die, we know. It's putting off*
> *The time of death that's of concern to men.*

Not being Shakespeare, I can't find a comment worthy of this, this project. I am tempted to say in an Elizabethan voice: "Ah, wayward Waypen, lascivious Layton, lay down thine errant pen!" And yet if I said this to them they would only reply earnestly, uncomprehendingly, sorrowfully: "Can't you give us some *con*structive criticism, not *de*-structive? Why don't you say *your* errant pen, not *thine?* And *lascivious!* Mr. Jarrell, if you *have* to talk about that type subject, don't say *lascivious* Layton, say *sexy* Layton!"

Even Little Red Riding Hood is getting too hard for children, I read. The headline of the story is CHILD'S BOOKS BEING MADE MORE SIMPLE; the story comes from New York, is distributed by the

International News Service, and is written by Miss Olga Curtis. Miss Curtis has interviewed Julius Kushner, the head of a firm that has been publishing children's books longer than anyone else in the country. He tells Miss Curtis:

"Non-essential details have disappeared from the 1953 Little Red Riding Hood story. Modern children enjoy their stories better stripped down to basic plot—for instance, Little Red Riding Hood meets wolf, Little Red Riding Hood escapes wolf. [I have a comment: the name Little Red Riding Hood seems to me both long and non-essential—why not call the child Red, and strip the story down to Red meets wolf, Red escapes wolf? At this rate, one could tell a child all of Grimm's tales between dinner and bed-time.]

" 'We have to keep up with the mood of each generation,' Kushner explained. 'Today's children like stories condensed to essentials, and with visual and tactile appeal as well as interesting content.'

"Modernizing old favorites, Kushner said, is fundamentally a matter of simplifying. Kushner added that today's children's books are intended to be activity games as well as reading matter. He mentioned books that make noises when pressed, and books with pop-up three-dimensional illustrations as examples of publishers' efforts to make each book a teaching aid as well as a story."

As one reads one sees before one, as if in a vision, the children's book of the future: a book that, pressed, says: *I'm your friend;* teaches the child that Crime Does Not Pay; does not exceed thirty words; can be used as a heating-pad if the electric blanket breaks down; and has three-dimensional illustrations dyed with harmless vegetable coloring matter and flavored with pure vanilla. I can hear the children of the future crying: "Mother, read us another vanilla book!"

But by this time you must be thinking, as I am, of one of the more frightening things about our age: that much of the body of common knowledge that educated people (and many uneducated people) once had, has disappeared or is rapidly disappearing. Fairy tales, myths, proverbs, history—the Bible and Shakespeare and Dickens, the *Odyssey* and *Gulliver's Travels*—these and all the things like them are surprisingly often things that most of an audience won't understand an allusion to, a joke about. These things were the ground on which the people of the past came together. Much of the wit or charm or elevation of any writing or conversation with an atmosphere depends upon this presupposed, easily and affectionately remembered body of common knowledge; because of it we understand things, feel about things, as human beings and not as human animals.

Who teaches us all this? Our families, our friends, our schools, society in general. Most of all, we hope, our schools. When I say *schools* I mean grammar schools and high schools and colleges—but the first two are more important. Most people still don't go to college, and those who do don't get there until they are seventeen or eighteen. "Give us a child until he is seven and he is ours," a Jesuit is supposed to have said; the grammar schools and high schools of the United States have a child for ten years longer, and then he is—whose? Shakespeare's? Leroy S. Layton's? The *Readers' Digest*'s? When students at last leave high school or go on to college, what are they like?

III

College teachers continually complain about their students' "lack of preparation," just as, each winter, they complain about the winter's lack of snow. Winters don't have as much snow as winters used to have: things are going to the dogs and always have been. The teachers tell one another stories about The Things Their Students Don't Know —it surprises you, after a few thousand such stories, that the students manage to find their way to the college. And yet, I have to admit, I have as many stories as the rest; and, veteran of such conversations as I am, I am continually being astonished at the things my students don't know.

One dark, cold, rainy night—the sort of night on which clients came to Sherlock Holmes—I read in a magazine that winters don't have as much snow as winters used to have; according to meteorologists, the climate *is* changing. Maybe the students are changing too. One is always hearing how much worse, or how much better, schools are than they used to be. But one isn't any longer going to grammar school, or to high school either; one isn't, like Arnold, a school inspector; whether one believes or disbelieves, blames or praises, how little one has to go on! Hearing one child say to another: "What does E come after in the alphabet?" makes a great, and perhaps unfair, impression on one. The child may not be what is called a random sample.

Sitting in my living room by the nice warm fire, and occasionally looking with pleasure at the rain and night outside—how glad I was that I wasn't in them!—I thought of some other samples I had seen just that winter, and I wasn't sure whether they were random, either. That winter I had had occasion to talk with some fifth-grade students and some eighth-grade students; I had gone to a class of theirs; I had even gone carolling, in a truck, with some Girl Scouts and their Scoutmistress, and had been dismayed at all the carols I didn't know—it was a part of my education that had been neglected.

I was not dismayed at the things the children hadn't known, I was overawed; there were very few parts of their education that had not been neglected. Half the fifth-grade children—you won't, just as I couldn't, believe this—didn't know who Jonah was; only a few had ever heard of King Arthur. When I asked an eighth-grade student about King Arthur she laughed at my question, and said: "Of course I know who King Arthur was." My heart warmed to her *of course*. But she didn't know who Lancelot was, didn't know who Guinevere was;

she had never heard of Sir Galahad. I realized with a pang the truth of the line of poetry that speaks of "those familiar, now unfamiliar knights that sought the Grail." I left the Knights of the Round Table for history: she didn't know who Charlemagne was.

She didn't know who Charlemagne was! And she had never heard of Alexander the Great; her class had "had Rome," but she didn't remember anything about Julius Caesar, though she knew his name. I asked her about Hector and Achilles: she had heard the name Hector, but didn't know who he was; she had never heard of "that other one."

I remembered the college freshman who, when I had asked her about "They that take the sword shall perish with the sword," had answered: "It's Shakespeare, I think"; and the rest of the class hadn't even known it was Shakespeare. Nobody in the class had known the difference between faith and works. And how shocked they had all been—the Presbyterians especially—at the notion of predestination!

But all these, except for the question of where E comes in the alphabet, had been questions of literature, theology, and European history; maybe there *are* more important things for students to know. The little girl who didn't know who Charlemagne was had been taught, I found, to conduct a meeting, to nominate, and to second nominations; she had been taught—I thought this, though far-fetched, truly imaginative—the right sort of story to tell an eighteen-months-old baby; and she had learned in her Domestic Science class to bake a date-pudding, to make a dirndl skirt, and from the remnants of the cloth to make a drawstring carryall. She could not tell me who Charlemagne was, it is true, but if I were an eighteen-months-old baby I could go to her and be sure of having her tell me the right sort of story. I felt a senseless depression at this; and thought, to alleviate it, of the date-pudding she would be able to bake me.

I said to myself about her, as I was getting into the habit of saying about each new eighth-grade girl I talked to: "She must be an exception"; pretty soon I was saying: "She *must* be an exception!" If I had said this to her teacher she would have replied: "Exception indeed! She's a nice, normal, well-adjusted girl. She's one of the drum-majorettes and she's Vice-President of the Student Body; she's had two short stories in the school magazine and she made her own formal for the Sadie Hawkins dance. She's an *exceptionally* normal girl!" And what could I have answered? "But she doesn't know who Charlemagne was"? You can see how ridiculous that would have sounded.

How many people cared whether or not she knew who Charlemagne was? How much good would knowing who Charlemagne was ever do her? Could you make a dirndl out of Charlemagne? make, even, a drawstring carryall? There was a chance—one chance in a hundred million—that someday, on a quiz program on the radio, someone would ask her who Charlemagne was. If she knew the audience would applaud in wonder, and the announcer would give her a refrigerator; if she didn't know the audience would groan in sympathy, and the announcer would give her a dozen cartons of soap-powder. Euclid, I believe, once gave a penny to a student who asked: "What good will studying geometry do me?"—studying geometry made *him* a penny. But knowing who Charlemagne was would in all probability never make her a penny.

Another of the eighth-grade girls had shown me her Reader. All the eighth-grade students of several states use it; it is named *Adventures for Readers.* It has in it, just as Readers used to have in them, *The Man Without a Country* and *The Legend of Sleepy Hollow* and *Evangeline*, and the preface to *Adventures for Readers* says about their being there: "The competition of movies and radios has reduced the time young children spend with books. It is no longer supposed, as it once was, that reading skills are fully developed at the end of the sixth grade . . . Included are *The Man Without a Country*, *The Legend of Sleepy Hollow*, and *Evangeline*. These longer selections were once in every eighth grade reading book. They have disappeared because in the original they are far too difficult for eighth grade readers. Yet, they are never presented for other years. If they are not read in the eighth grade, they are not read at all. In their simplified form they are once more available to young people to become a part of their background and experience."

I thought that in the next edition of *Adventures for Readers* the editors would have to substitute for the phrase *the competition of movies and radios,* the phrase *the competition of movies, radios, and television:* I thought of this thought for some time. But when I thought of Longfellow's being *in the original* far too difficult for eighth-grade students, I—I did not know what to think. How much more difficult everything is than it used to be!

I remembered a letter, one about difficult writers, that I had read in *The Saturday Review.* The letter said: "I have been wondering when somebody with an established reputation in the field of letters would stand tiptoe and slap these unintelligibles in the face. Now I hope the publishers will wake up and throw the whole egotistical, sophist lot of them down the drain. I hope that fifty years from now nobody will remember that Joyce or Stein or James or Proust or Mann ever lived."

I knew that such feelings are not peculiar to our own place or age. Once while looking at an exhibition of Victorian posters and paintings and newspapers and needlework, I had read a page of the London *Times,* printed in the year 1851, that had on it a review of a new book by Alfred Tennyson. After several sentences about what was wrong with this book, the reviewer said: "Another fault is not particular to *In Memoriam;* it runs through all Mr.

Tennyson's poetry—we allude to his obscurity." And yet the reviewer would not have alluded to Longfellow's obscurity; those Victorians for whom everything else was too difficult still understood and delighted in Longfellow. But Tennyson had been too obscure for some of them, just as Longfellow was getting to be too obscure for some of us, as our "reading skills" got less and less "fully developed."

This better-humored writer of the London *Times* had not hoped that in fifty years nobody would remember that Tennyson had ever lived; and this is fortunate, since he would not have got his wish. But I thought that the writer to *The Saturday Review* might well get, might already be getting, a part of his wish. How many people there were all around him who did not remember—who indeed had never learned—that Proust or James or Mann or Joyce had ever lived! How many of them there were, and how many more of their children there were, who did not remember—who indeed had never learned—that Jonah or King Arthur or Galahad or Charlemagne had ever lived! And in the end all of us would die, and not know, then, that anybody had ever lived: and the writer to *The Saturday Review* would have got not part of his wish but all of it.

And if, in the meantime, some people grieved to think of so much gone and so much more to go, they were the exception. Or, rather, the exceptions: millions and millions—tens of millions, even—of exceptions. There were enough exceptions to make a good-sized country; I thought, with pleasure, of walking through the streets of that country and having the children tell me who Charlemagne was.

I decided not to think of Charlemagne any more, and turned my eyes from my absurd vision of the white-bearded king trying to learn to read, running his big finger slowly along under the words . . . My samples weren't really random, I knew; I was letting myself go, being exceptionally unjust to that exceptionally normal girl and the school that had helped to make her so. She was being given an education suitable for the world she was to use it in; my quarrel was not so much with her education as with her world, and our quarrels with the world are like our quarrels with God: no matter how right we are, we are wrong. But who wants to be right all the time? I thought, smiling; and said goodbye to Charlemagne with the same smile.

Instead of thinking, I looked at *The New York Times Book Review;* there in the midst of so many books, I could surely forget that some people don't read any. And after all, as Rilke says in one of his books, we are—some of us are—*beaten at/By books as if by perpetual bells;* we can well, as he bids us, *rejoice/When between two books the sky shines to you, silent.* In the beginning was the Word, and man has made books of it.

I read quietly along, but the review I was reading was continued on page 47; and as I was turning to page 47 I came to an advertisement, a two-page advertisement of the Revised Standard Version of the Bible. It was a sober, careful, authorized sort of advertisement, with many testimonials of clergymen, but it was, truly, an advertisement. It said:

"In these anxious days, the Bible offers a practical antidote for sorrow, cynicism, and despair. But the King James version is often difficult reading.

"If *you* have too seldom opened your Bible because the way it is written makes it hard for you to understand, the Revised Standard Version can bring you an exciting new experience.

"Here is a Bible so enjoyable you find you pick it up twice as often . . ."

Tennyson and Longfellow and the Bible—what *was* there that wasn't difficult reading? And a few days before that I had torn out of the paper—I got it and read it again, and it was hard for me to read it—a Gallup Poll that began: "Although the United States has the highest level of formal education in the world, fewer people here read books than in any other major democracy." It didn't compare us with minor autocracies, which are probably a lot worse. It went on to say that "fewer than one adult American in every five was found to be reading a book at the time of the survey. [Twenty years ago, 29% were found to be reading a book; today only 17% are.] In England, where the typical citizen has far less formal schooling than the typical citizen here, nearly three times as many read books. Even among American college graduates fewer than half read books."

It went on and on; I was so tired that, as I read,

the phrase *read books* kept beating in my brain, and getting mixed up with Charlemagne: compared to other major monarchs, I thought sleepily, fewer than one-fifth of Charlemagne reads books. I read on as best as I could, but I thought of the preface to *Adventures for Readers,* and the letter to *The Saturday Review,* and the advertisement in *The New York Times Book Review,* and the highest level of formal education in the world, and they all went around and around in my head and said to me an advertisement named *Adventures for Non-Readers:*

"In these anxious days, reading books offers a practical antidote to sorrow, cynicism, and despair. But books are often, in the original, difficult reading.

"If *you* have too seldom opened books because the way they are written makes them hard for you to understand, our Revised Standard Versions of books, in their simplified, televised form, can bring you an exciting new experience.

"Here are books so enjoyable you find you turn them on twice as often."

I shook myself; I was dreaming. As I went to bed the words of the eighth-grade class's teacher, when the class got to *Evangeline,* kept echoing in my ears: "We're coming to a long poem now, boys and girls. Now don't be babies and start counting the pages." I lay there like a baby, counting the pages over and over, counting the pages.

The Schools of Yesteryear (A One-Sided Dialogue)

UNCLE WADSWORTH (*a deep, slightly grained or corrugated, comfortable-sounding voice, accompanied by an accordion*): School days, school days, dear old golden rule—

ALVIN (*Alvin's voice is young*): Stop, Uncle Wadsworth!

UNCLE WADSWORTH: Why should I stop, Alvin boy?

ALVIN: Because it isn't *so*, Uncle Wadsworth. (*With scorn.*) Dear old golden rule days! That's just nostalgia, just sentimentality. The man that wrote that song was just too old to remember what it was really like. Why, kids hated school in those days—they used to play hookey all the time. It's different now. Children *like* to go to school now.

UNCLE WADSWORTH: Finished, Alvin boy?

ALVIN: Finished, Uncle Wadsworth.

UNCLE WADSWORTH: School days, school days, dear old golden rule days, Readin' and 'ritin' and 'rithmetic, Taught to—

ALVIN: Stop, Uncle Wadsworth!

UNCLE WADSWORTH: Why should I stop this time, Alvin boy?

ALVIN: Reading and writing and arithmetic! What a curriculum! Why, it sounds like it was invented by an illiterate. How could a curriculum like that prepare you for life? No civics, no social studies, no hygiene; no home economics, no manual training, no physical education! And extra-curricular activities—where were they?

UNCLE WADSWORTH: Where indeed? Where are the extra-curricular activities of yesteryear? Shall I go on, Alvin boy?

ALVIN: Go ahead, Uncle Wadsworth.

UNCLE WADSWORTH: School days, school days, dear old golden rule days, Readin' and 'ritin' and 'rithmetic, Taught to the tune of a hick'ry stick—

ALVIN: Stop! Stop! Stop, Uncle Wadsworth! (*He pants with emotion.*) Honestly, Uncle, I don't see how you can bear to say it. *Taught to the tune of a hickory stick!* . . . Imagine having to *beat* poor little children with a *stick!* Thank God those dark old days of ignorance and fear and compulsion are over, and we just appeal to the child's better nature, and get him to adjust, and try to make him see that what he likes to do is what we want him to do.

UNCLE WADSWORTH: Finished, Alvin boy?

ALVIN: Finished, Uncle Wadsworth.

UNCLE WADSWORTH: Well, so am I. I can't seem to

get going in this song—every fifty yards I get a puncture and have to stop for air. You go on for a while and let me interrupt you. Go ahead, Alvin.

ALVIN: Go ahead where?

UNCLE WADSWORTH: Go ahead about those dark old days of ignorance and fear and compulsion. It makes my flesh creep—and I'm just like the fat boy, I *like* to have my flesh creep.

ALVIN: What fat boy?

UNCLE WADSWORTH: The one in *Pickwick Papers.* (*Silence from Alvin.*) You know, *Pickwick Papers.* (*Silence from Alvin.*) It's a book, son—a book by Charles Dickens. Ever read any Dickens?

ALVIN: Oh, sure, sure. I read *The Tale of Two Cities* in high school. And *Oliver Twist*—well, really I didn't read it exactly, I read it in *Illustrated Classics.* And I saw *Great Expectations* in the movies.

UNCLE WADSWORTH: Why, you and Dickens are old friends. But go on about the—the schools of yesteryear.

ALVIN: Well, I will, Uncle Wadsworth. After all, it's only because I was lucky enough to be born now that I didn't have to go to one of those schools myself. I can just see myself trudging to school barefooted in my overalls—because they didn't even have school buses then, you know—

UNCLE WADSWORTH: Not a one! If a school bus had

come for me I'd have thought it was a patrol wagon someone had painted orange for Hallowe'en.

ALVIN: Well, there I am trudging along, and I'm not only trudging, I'm *limping.*

UNCLE WADSWORTH: Stub your toe?

ALVIN (*with bitter irony*): Stub my toe! I'm limping because I'm *sore*—sore all over, where the teacher beat me.

UNCLE WADSWORTH: All over isn't where the teacher beat you, Alvin boy—I know.

ALVIN: All right, all right! And when I get to the school is it the Consolidated School? Is there a lunch-room and a 'chute-the-'chute and a jungle-gym? Is it—is it like schools ought to be? Uh-*uh!* That school has one room, and it's *red.*

UNCLE WADSWORTH: You mean that even in those days the Communists—

ALVIN: No, no, not Red, *red!* Red like a barn. And when I get inside, the teacher is an old maid that looks like a broomstick, or else a man that looks like a—that looks like Ichabod Crane. And then this Crane-type teacher says to me, real stern: "Alvin McKinley, stand up! Are you aware, Alvin, that it is *three minutes past seven?*"

UNCLE WADSWORTH: Three minutes past seven! What on earth are you and Ichabod Crane doing in school at that ungodly hour?

ALVIN: That's when school starts then! Or six, maybe. . . . Then he says, pointing his finger at me in a terrible voice: "Three minutes tardy!

And what, Alvin, what did I warn you would happen to you if you ever again were so much as one minute tardy? What did I tell you that I would do to you?" And I say in a little meek voice, because I'm scared, I say: "Whip me." And he says: "YES, WHIP YOU!" And I say—

UNCLE WADSWORTH: You say, "Oh, *don't* whip pore Uncle Tom, massa! If only you won't whip him he won't never—"

ALVIN: Oh, stop it, Uncle Wadsworth! That's not what I say at all, and you know it. How can I tell about the schools of yesteryear if you won't be serious? Well, anyway, he says to me: "Have you anything to say for yourself?" And I say, "Please, Mr. Crane, it was four miles, and I had the cows to milk, and Ma was sick and I had to help Sister cook the hoe-cakes—"

UNCLE WADSWORTH: Hoe-cakes! (*With slow relish.*) Hoe-cakes. . . . Why, I haven't had any hoe-cakes since. . . . How'd you hear about hoe-cakes, Alvin boy?

ALVIN: Uncle Wadsworth, if you keep interrupting me about irrevu—irrelevancies, how can I get anywhere?

UNCLE WADSWORTH: I apologize, Alvin; I am silent, Alvin.

ALVIN: Then he looks at me and he smiles like—like somebody in *Dick Tracy*, and he says: "Alvin, *spare your breath.*" And then he walks over to the corner next to the stove, and do you know what's in the corner?

UNCLE WADSWORTH: What's in the corner?

ALVIN: Sticks. Sticks of every size. Hundreds of sticks. And then he reaches down and takes the biggest one and—and—

UNCLE WADSWORTH: And—and—

ALVIN: And he *beats* me.

UNCLE WADSWORTH (*with a big sigh*): The Lord be praised! For a minute I was afraid he was going to burn you at the stake. But go ahead, Alvin.

ALVIN: Go ahead?

UNCLE WADSWORTH: It's still just ten minutes after seven. Tell me about your day—your school-day—your dear old golden rule day.

ALVIN: Well, then he says: "Take your Readers!" And I look around and everybody in the room, from little kids just six years old with their front teeth out to great big ones, grown men practically that look like they ought to be on the Chicago Bears—everybody in the room picks up the same book and they all start reading aloud out of the—*McGuffey Reader!* Ha-ha-ha! The McGuffey Reader!

UNCLE WADSWORTH: And why, Alvin, do you laugh?

ALVIN: Because it's funny, that's why! The McGuffey Reader!

UNCLE WADSWORTH: Have you ever seen a McGuffey Reader, Alvin?

ALVIN: How could I of, Uncle Wadsworth? I didn't go to school back in those days.

UNCLE WADSWORTH: Your account was so vivid that for a moment I forgot. . . . You've never seen such a Reader. Well, I have.

ALVIN: Oh, sure—you used one in school yourself, didn't you?

UNCLE WADSWORTH: No, Alvin—strange as it seems, I did not; nor did I ever shake the hand of Robert E. Lee, nor did I fight in the War of 1812, nor did I get to see Adam and Eve and the Serpent. My father used a McGuffey Reader; I did not.

ALVIN: I'm sorry, Uncle Wadsworth.

UNCLE WADSWORTH: No need, no need. . . . Alvin, if you will go over to the bookcase and reach into the right hand corner of the top shelf, you will find a book—a faded, dusty, red-brown book.

ALVIN: Here it is. It's all worn, and there're gold letters on the back, and it says *Appletons' Fifth Reader*.

UNCLE WADSWORTH: Exactly. *Appletons' Fifth Reader*. Week before last, at an antique-dealer's over near Hillsboro, side by side with a glass brandy-flask bearing the features of the Father of our Country, George Washington, I found this Reader.

ALVIN: Look how yellow the paper is! And brown spots all over it. . . . Gee, they must have used it all over the country; it says New York, Boston, and Chicago, 1880, and it was printed in 1878 and 1879 too, and—look at the picture across from it, it's one of those old engravings. I guess they didn't have photographs in those days.

UNCLE WADSWORTH: Guess again, Alvin boy. And what is the subject of this old engraving?

ALVIN: A girl with a bucket, and back behind her somebody's plowing, and it's dawn. And there's some poetry underneath.

UNCLE WADSWORTH:

While the plowman near at hand
Whistles o'er the furrowed land
And the milkmaid singeth blithe. . . .

ALVIN: That's right! You mean to say you *memorized* it?

UNCLE WADSWORTH: Fifty years ago, Alvin. Doesn't any of it have a—a familiar ring?

ALVIN: Well, to tell the truth, Uncle Wadsworth. . . .

UNCLE WADSWORTH: What does it say in small letters down at the right-hand corner of the page?

ALVIN: It says—"*L'Allegro*, page 420." *L'Allegro!* Sure! sure! Why, I read it in sophomore English. We spent two whole days on that poem and on —you know, that other one that goes with it. They're by John Milton.

UNCLE WADSWORTH: Yes, Milton. And in that same—

ALVIN: But Uncle Wadsworth, you don't mean to say they had Milton in a Fifth Reader! Why, we were sophomores in college, and there were two football players that were juniors, and believe me, it was all Dr. Taylor could do to get us through that poem. How could little kids in the fifth grade read Milton?

UNCLE WADSWORTH: Sit down, Alvin. Do you remember reading, at about the same time you read "L'Allegro," a poem called "Elegy Written in a Country Churchyard"?

ALVIN: Well—

UNCLE WADSWORTH: Gray's "Elegy"?

ALVIN: Say me some, Uncle Wadsworth.

UNCLE WADSWORTH:

Full many a gem of purest ray serene
The dark unfathom'd caves of ocean bear;
Full many a flower is born to blush unseen
And waste its sweetness on the desert air.

ALVIN: Sure, I remember that one. I liked that one.

UNCLE WADSWORTH: Well, Alvin, that very poem—

ALVIN: Oh *no*, Uncle Wadsworth! You're not going to tell me that that poem was in a Fifth Reader!

UNCLE WADSWORTH: No, Alvin, I am not. I want you to . . . to steel yourself. That poem was not in Appletons' Fifth Reader, that poem was in Appletons' Fourth Reader. (*Alvin groans in awe.*) And Wordsworth—you studied Wordsworth in your sophomore English?

ALVIN (*lifelessly*): Uh-huh.

UNCLE WADSWORTH: There are four of Wordsworth's poems in Appletons' Fourth Reader.

ALVIN: I guess in the Sixth Reader they were reading Einstein.

UNCLE WADSWORTH: No, but in the Fifth Reader—

run your eye down the table of contents, Alvin—there are selections by Addison, Bishop Berkeley, Bunyan, Byron, Coleridge, Carlyle, Cervantes, Coleridge—the whole *Ancient Mariner*, Alvin—Defoe, De Quincy, Dickens, Emerson, Fielding, Hawthorne, George Herbert, Hazlitt, Jefferson, Dr. Johnson, Shakespeare, Shelley, Sterne, Swift, Tennyson, Thoreau, Mark Twain—

ALVIN: It's hard to believe.

UNCLE WADSWORTH: And there are also selections from simpler writers—

ALVIN: Yeah, simple ones—

UNCLE WADSWORTH: Simpler writers such as Scott, Burns, Longfellow, Cooper, Audubon, Poe, Oliver Wendell Holmes, Benjamin Franklin, Washington Irving. Alvin, have you ever—at college perhaps—ever read anything by Goethe?

ALVIN: I don't *believe* so.

UNCLE WADSWORTH: Well, Alvin boy, if after milking the cow and baking the hoe-cakes, you had limped four miles barefoot to that one-room red schoolhouse of yours, and had been beaten by that Ichabod Crane of a teacher, you would still have got to read, in your Appletons' Fifth Reader, one poem and five pages of prose from Goethe's immortal *Wilheim Meister*. . . . As it is you don't limp, nobody beats you, and you read—whom *do* you read, Alvin? Tell me some of the writers you read in the fifth grade.

ALVIN: I don't exactly remember their *names*.

UNCLE WADSWORTH: There in the bookcase—that red and yellow and black book there—is the Fifth Reader of today. *Days and Deeds*, it is called; it is, I believe, the most popular Fifth Reader in the country. That's right, hand it over. Here on page 3 is its table of contents; come, Alvin, read out to me the names of the writers from whom the children of today get their knowledge of life and literature.

ALVIN: Well, the first one's Fletcher D. Slater, and then Nora Burglon, and Sterling North and Ruth G. Plowhead—

UNCLE WADSWORTH: Plowhead?

ALVIN: That's what it says. Then Ruth E. Kennell, Gertrude Robinson, Philip A. Rollins, J. Walker McSpadden, Merlin M.—

UNCLE WADSWORTH: You're sure you're not making up some of these names?

ALVIN: How could I? Merlin M. Taylor, Sanford Tousey, Gladys M. Wick, Marie Barton, Margaret Leighton, Edward C. James—no, Janes, Leonard K. Smith, P. L. Travers, Esther Shepherd, James C. Bowman, Dr. Seuss—

UNCLE WADSWORTH: Land! Land!

ALVIN: No, Seuss. Seuss.

UNCLE WADSWORTH: I speak figuratively. I mean that here, at last, is a name I recognize, the name of a well-known humorist and cartoonist.

ALVIN: Oh. Then there's Armstrong Sperry, Myra M. Dodds, Alden G. Stevens, Lavinia R. Davis, Lucy M. Crockett, Raymond Jannson,

Hubert Evans, Ruth E. Tanner, Three Boy Scouts—

UNCLE WADSWORTH: Three Boy Scouts. An Indian, no doubt. . . . Never heard of him.

ALVIN: Heard of *them.* There're three of them.

UNCLE WADSWORTH: Three? Thirty! Three hundred! They're *all* Boy Scouts! Alvin, these are names only a mother could love—names only a mother would know. That they are honest names, respected names, the names of worthy citizens, I have not the slightest doubt; but when I reflect that it is *these* names that have replaced those of Goethe, of Shakespeare, of Cervantes, of Dr. Johnson—of all the other great and good writers of the Appleton Fifth Reader—when I think of this, Alvin, I am confused, I am dismayed, I am *astounded.*

ALVIN: Uncle Wadsworth, you've got all red in the face.

UNCLE WADSWORTH: There are also in the Appleton Fifth Reader, Alvin, elaborate analyses of the style, rhetoric, and organization of the literary works included; penetrating discussions of their logic; highly technical instructions for reading them aloud in the most effective way; discussions of etymology, spelling, pronunciation, the general development of the English language. And, Alvin, these are *not* written in the insipid baby-talk thought appropriate for children today. Here, in a paragraph about *Don Quixote,* is one of the Fifth Reader's typical discussions of logic: "The question here involved is the old sophism of Eubulides. . . . Is a man a liar who says that he tells lies? If he is, then he does not tell lies; and if he does not tell lies, is he a liar? If not, then is not his assertion a lie? . . . It will be noticed that the perplexity comes from the fact of self-relation: the one assertion relates to another assertion of the same person; and the one assertion being conditioned upon the other, the difficulty arises. It is the question of self-contradiction—of two mutually contradictory statements, one must be false. It is a sophism, but one that continually occurs among unsophisticated reasoners. It is also a practical sophism, for it is continually being acted in the world around us (e.g., a person seeks pleasure by such means that, while he enjoys himself, he undermines his health, or sins against his conscience, and thus draws inevitably on him physical suffering and an uneasy soul). It is therefore well worthy of study in its purely logical form. . . . All universal negative assertions (and a lie is a negation) are liable to involve the assertion itself in self-contradiction."

ALVIN: Ohhhhh. . . . *Ohhhhh*. . . . If I'd gone to school then, I'd have known what that means in the *fifth grade?*

UNCLE WADSWORTH: You'd have known it or you never would have got into the sixth grade.

ALVIN: Then I'd be the oldest settler in the fifth grade, because I'm a junior in college and I still can't understand it.

UNCLE WADSWORTH: Yes, it's surprising what those fifth-graders were expected to know. The Reader contains a little essay called "Hidden Beauties of Classic Authors," by a writer named N. P. Willis.

ALVIN: N. P. Willis. . . . I guess he was Ruth G. Plowhead's grandpa.

UNCLE WADSWORTH: Yes, he isn't exactly a classic author himself. He tells you how he fell in love with Beaumont and Fletcher, and the *Faerie Queene*, and *Comus*, and *The Rape of the Lock;* he says that he knows "no more exquisite sensation than this warming of the heart to an old author; and it seems to me that the most delicious portion of intellectual existence is the brief period in which, one by one, the great minds of old are admitted with all their time-mellowed worth to the affections." Well, at the end of the essay there're some questions; what do you think is the first thing they ask those fifth-graders?

ALVIN: What?

UNCLE WADSWORTH: "Have you read Milton's *Comus?*—Pope's *Rape of the Lock?*"

ALVIN: Now Uncle Wadsworth, you've got to admit that that's a terrible thing to ask a little boy in the fifth grade.

UNCLE WADSWORTH: *I* think it's a terrible thing. But they didn't. As a matter of fact, *I* think it's a terrible thing to ask a big boy in his junior year in college. How about it, Alvin? Have *you* read Milton's *Comus?* Pope's *Rape of the Lock?*

ALVIN: Well, to tell you the truth, Uncle Wadsworth—

UNCLE WADSWORTH: Tell ahead.

ALVIN: Well, to—well—well, it just isn't the *sort* of question you can answer yes or no. I *may* have read Milton's *Comus;* it's the kind of thing we read hundreds of things like in our sophomore survey course; I guess the chances are ten to one I read it, and a year ago I could have told you for certain whether or not I read it, but right now all I can say is if I didn't read it, it would surprise me a lot.

UNCLE WADSWORTH: And *The Rape of the Lock?*

ALVIN: No.

UNCLE WADSWORTH: *No?* You mean you *know* you didn't read it?

ALVIN: Uh-huh.

UNCLE WADSWORTH: How do you know?

ALVIN: I—

UNCLE WADSWORTH: Go on, go on.

ALVIN: Well Uncle Wadsworth, it seems to me that a book with a title like that, if I'd read it I'd remember it.

UNCLE WADSWORTH: Alvin, if you weren't my own nephew I'd—I'd be proud to have invented you. . . . Here's another of those poems, the kind that *you* read in your sophomore year in college and that your great-grandfather read in the Fifth Reader. It's by George Herbert, the great rēligious poet George Herbert. Read it to me, Alvin; and when you've read it, tell me what it means.

ALVIN (*in careful singsong*): *Sunday*. By George Herbert.

O Day most calm, most bright!
The fruit of this, the next world's bud;
The endorsement of supreme delight,
Writ by a Friend, and with his blood;
The couch of Time: Care's calm and bay:
The week were dark but for thy light;
Thy torch doth show the way.

The other days and thou
Make up one man, whose face thou art,
Knocking at heaven with thy brow:
The working-days are the back part;
The burden of the week lies there;
Making the whole to stoop and bow,
Till thy release appear.

Man had—man had—

Uncle Wadsworth, I'm all mixed up. I've *been* all mixed up. And if you ask me that fifth grade was mixed up too.

UNCLE WADSWORTH: Where did you first begin to feel confused?

ALVIN: I never did not feel confused.

UNCLE WADSWORTH: Surely the first line—

ALVIN: Yeah. Yeah. The first one was all right. *O Day most calm, most bright!* That means it's Sunday, and it's all calm and bright, the weather's all calm and bright. Then it says, *the fruit of this*. . . . *The fruit of this*. What's the fruit of this?

UNCLE WADSWORTH: *The fruit of this, the next world's bud. World* is understood.

ALVIN: Understood?

UNCLE WADSWORTH: Yes. The fruit of this world and the bud of the next world.

ALVIN: Oh. . . . *The endorsement of supreme delight.* (*Pauses.*) *The endorsement of supreme delight.* . . . Uncle Wadsworth, a line like that— you've got to admit a line like that's *obscure.*

UNCLE WADSWORTH: It means that—it *says* that Sunday is like the endorsement of a check or note; because of the endorsement this supreme delight, our salvation, is negotiable, we can cash it.

ALVIN: Oh. . . . Like endorsing a check. *Writ by a Friend—Friend's* got a capital *F.* . . . Oh! That means it was written by a Quaker. (*Uncle Wadsworth laughs.*) But that's what it does mean. We live on a road named the Friendly Road because it goes to a Quaker church. If *Friend* doesn't mean *Quaker* why's it got a capital *F?*

UNCLE WADSWORTH: *Writ by a Friend, and with his blood.* If you're talking about church and Sunday and the next world, and mention a Friend who has written something with his blood, who is that Friend, Alvin?

ALVIN: Oh. . . . *The couch of Time; Care's calm and bay.* . . . (*Pauses.*) Uncle Wadsworth, do we *have* to read poetry?

UNCLE WADSWORTH: Of course not, Alvin. Nobody

else does, why should we? Let's get back to prose. Here's the way the Fifth Reader talks about climbing a mountain: "Some part of the beholder, even some vital part, seems to escape through the loose grating of his ribs as he ascends. . . . He is more lone than you can imagine. There is less of substantial thought and fair understanding in him than in the plains where men inhabit. His reason is dispersed and shadowy, more thin and subtle, like the air. Vast, Titanic, inhuman Nature has got him at disadvantage, caught him alone, and pilfers him of some of his divine faculty. She does not smile on him as in the plains. She seems to say sternly, 'Why come ye here before your time? . . . Why seek me where I have not called you, and then complain because you find me but a stepmother? Shouldst thou freeze, or starve, or shudder thy life away, here is no shrine, nor altar, nor any access to my ear. "Chaos and ancient Night, I come no spy/ With purpose to explore or to disturb/ The secrets of your realm—"' "

ALVIN: Uncle Wadsworth, if the prose is like that, I'd just as soon have stayed with the poetry. Didn't they have any plain American writers in that Fifth Reader?

UNCLE WADSWORTH: Plain American writers? That was Thoreau I was reading you. Well, if he's too hard, here's what the Fifth Reader has to say about him. It's talking about his account of the battle between the black ants and the red: "The style of this piece is an imitation of the heroic style of Homer's 'Iliad,' and is properly a 'mock-heroic.' The intention of the author is two-fold: half-seriously endowing the incidents of every-day life with epic dignity, in the belief that there is nothing mean and trivial to the poet and philosopher, and that it is the man that adds dignity to the occasion, and not the occasion that dignifies the man; half-satirically treating the human events alluded to as though they were non-heroic, and only fit to be applied to the events of animal life."

ALVIN (*wonderingly*): Why, it's just like old Taylor!

UNCLE WADSWORTH: Professor Taylor would lecture to you in that style?

ALVIN: He'd get going that way, but pretty soon he'd see we didn't know what he meant, and then he'd talk so we could understand him. . . . Well, if the Fifth Reader sounds like that about ants, I sure don't want to hear it about scansion and etymology!

UNCLE WADSWORTH: But Alvin, wouldn't you *like* to be able to understand it? Don't you wish you'd had it in the fifth grade and known what it was talking about?

ALVIN: Sure, sure! Would I have made old Taylor's eyes pop out! All we ever had in the fifth grade was Boy Scouts going on hikes, and kids going to see their grandmother for Thanksgiving; it was easy.

UNCLE WADSWORTH: And interesting?

ALVIN: Nah, it was corny—the same old stuff; how can you make stuff like that interesting?

UNCLE WADSWORTH: How indeed?

ALVIN: But how did things like Shakespeare and Milton and Dickens ever get in a Fifth Reader?

UNCLE WADSWORTH: Alvin, they've always *been* there. Yesterday, here in the United States, those things were in the Fifth Reader; today, everywhere else in the world, those things or their equivalent are in the Fifth Reader; it is only here in the United States, today, that the Fifth Reader consists of *Josie's Home Run*, by Ruth G. Plowhead, and *A Midnight Lion Hunt*, by Three Boy Scouts. I read, in a recent best-seller, this sentence: "For the first time in history Americans see their children getting less education than they got themselves." That may be; and for the first time in history Americans see a book on why their children can't read becoming a best-seller, being serialized in newspapers across the nation. Alvin, about school-buildings, health, lunches, civic responsibility, kindness, good humor, spontaneity, we have nothing to learn from the schools of the past; but about reading, with pleasure and understanding, the best that has been thought and said in the world—about *that* we have much to learn. The child who reads and understands the Appleton Fifth Reader is well on the way to becoming an educated, cultivated human being—and if he has to do it sitting in a one-room schoolhouse, if he has to do it sitting on a hollow log, he's better off than a boy sitting in the Pentagon reading *Days and Deeds*. There's a jug of cider in the ice-box, Alvin; you get it, I'll get the glasses; and let's drink a toast to—

ALVIN: To the Appleton Fifth Reader! long may she read! (*They drink.*)

UNCLE WADSWORTH: And now, Alvin, let us conclude the meeting with a song.

ALVIN: What song?

UNCLE WADSWORTH: What song? Alvin, can you ask? Start us off, Alvin!

ALVIN: School days, school days. . . .

BOTH: Dear old golden rule days. . . .

ALVIN: Louder, Uncle Wadsworth, louder!

BOTH: Readin' and 'ritin' and 'rithmetic
Taught to the tune of a hick'ry stick. . . .

(*Alvin and Uncle Wadsworth and the accordion disappear into the distance.*)

A Sad Heart at the Supermarket

THE Emperor Augustus would sometimes say to his Senate: "Words fail me, my Lords; nothing I can say could possibly indicate the depth of my feelings in this matter." But in this matter of mass culture, the mass media, I am speaking not as an emperor but as a fool, a suffering, complaining, helplessly non-conforming poet-or-artist-of-a-sort, far off at the obsolescent rear of things; what I say will indicate the depth of my feelings and the shallowness and one-sidedness of my thoughts. If those English lyric poets who went mad during the eighteenth century had told you why the Age of Enlightenment was driving them crazy, it would have had a kind of documentary interest: what I say may have a kind of documentary interest. *The toad beneath the harrow knows/ Exactly where each tooth-point goes:* if you tell me that the field is being harrowed to grow grain for bread, and to create a world in which there will be no more famines, or toads either, I will say: "I know"; but let me tell you where the tooth-points go, and what the harrow looks like from below.

Advertising men, businessmen speak continually of *media* or *the media* or *the mass media.* One of their trade journals is named, simply, *Media.* It is an impressive word: one imagines Mephistopheles offering Faust *media that no man has ever known;* one feels, while the word is in one's ear, that abstract, overmastering powers, of a scale and intensity unimagined yesterday, are being offered one by the technicians who discovered and control them—offered, and at a price. The word has the clear fatal ring of that new world whose space we occupy so luxuriously and precariously; the world that produces mink stoles, rockabilly records, and tactical nuclear weapons by the million; the world that Attila, Galileo, Hansel and Gretel never knew.

And yet, it's only the plural of *medium.* "*Medium,*" says the dictionary, "that which lies in the middle; hence, middle condition or degree . . . A substance through which a force acts or an effect is transmitted . . . That through or by which anything is accomplished; as, an advertising *medium* . . . *Biol.* A nutritive mixture or substance, as broth, gelatin, agar, for cultivating bacteria, fungi, etc."

Let us name *our* trade journal *The Medium.* For all these media—television, radio, movies, newspapers, magazines, and the rest—are a single medium, in whose depths we are all being cultivated. This Medium is of middle condition or degree, mediocre;

it lies in the middle of everything, between a man and his neighbor, his wife, his child, his self; it, more than anything else, is the substance through which the forces of our society act upon us, and make us into what our society needs.

And what does it need? For us to need.

Oh, it needs for us to do or be many things: workers, technicians, executives, soldiers, housewives. But first of all, last of all, it needs for us to be buyers; consumers; beings who want much and will want more—who want consistently and insatiably. Find some spell to make us turn away from the stoles, the records, and the weapons, and our world will change into something to us unimaginable. Find some spell to make us see that the product or service that yesterday was an unthinkable luxury today is an inexorable necessity, and our world will go on. It is the Medium which casts this spell—which is this spell. As we look at the television set, listen to the radio, read the magazines, the frontier of necessity is always being pushed forward. The Medium shows us what our new needs are—how often, without it, we should not have known!—and it shows us how they can be satisfied: they can be satisfied by buying something. The act of buying something is at the root of our world; if anyone wishes to paint the genesis of things in our society, he will paint a picture of God holding out to Adam a check-book or credit card or Charge-A-Plate.

But how quickly our poor naked Adam is turned

into a consumer, is linked to others by the great chain of buying!

> *No outcast he, bewildered and depressed:*
> *Along his infant veins are interfused*
> *The gravitation and the filial bond*
> *Of nature that connect him with the world.*

Children of three or four can ask for a brand of cereal, sing some soap's commercial; by the time that they are twelve or thirteen they are not children but teen-age consumers, interviewed, graphed, analyzed. They are well on their way to becoming that ideal figure of our culture, the knowledgeable consumer. Let me define him: the knowledgeable consumer is someone who, when he comes to Weimar, knows how to buy a Weimaraner.

Daisy's voice sounded like money; everything about the knowledgeable consumer looks like or sounds like or feels like money, and informed money at that. To live is to consume, to understand life is to know what to consume: he has learned to understand this, so that his life is a series of choices —correct ones—among the products and services of the world. He is able to choose to consume something, of course, only because sometime, somewhere, he or someone else produced something—but just when or where or what no longer seems to us of as much interest. We may still go to Methodist or Baptist or Presbyterian churches on Sunday, but the Protestant ethic of frugal industry, of production for its own sake, is gone.

Production has come to seem to our society not

much more than a condition prior to consumption. "The challenge of today," an advertising agency writes, "is to make the consumer raise his level of demand." This challenge has been met: the Medium has found it easy to make its people feel the continually increasing lacks, the many specialized dissatisfactions (merging into one great dissatisfaction, temporarily assuaged by new purchases) that it needs for them to feel. When in some magazine we see the Medium at its most nearly perfect, we hardly know which half is entertaining and distracting us, which half making us buy: some advertisement may be more ingeniously entertaining than the text beside it, but it is the text which has made us long for a product more passionately. When one finishes *Holiday* or *Harper's Bazaar* or *House and Garden* or *The New Yorker* or *High Fidelity* or *Road and Track* or—but make your own list—buying something, going somewhere seems a necessary completion to the act of reading the magazine.

Reader, isn't buying or fantasy-buying an important part of your and my emotional life? (If you reply, *No*, I'll think of you with bitter envy as more than merely human; as deeply un-American.) It is a standard joke that when a woman is bored or sad she buys something, to cheer herself up; but in this respect we are all women together, and can hear complacently the reminder of how feminine this consumer-world of ours has become. One imagines as a characteristic dialogue of our time an interview in which someone is asking of a vague gracious figure, a kind of Mrs. America: "But while you waited for the intercontinental ballistic missiles what did you *do?*" She answers: "I bought things."

She reminds one of the sentinel at Pompeii—a space among ashes, now, but at his post: she too did what she was supposed to do. Our society has delivered us—most of us—from the bonds of necessity, so that we no longer struggle to find food to keep from starving, clothing and shelter to keep from freezing; yet if the ends for which we work and of which we dream are only clothes and restaurants and houses, possessions, consumption, how have we escaped?—we have exchanged man's old bondage for a new voluntary one. It is more than a figure of speech to say that the consumer is trained for his job of consuming as the factory-worker is trained for his job of producing; and the first can be a longer, more complicated training, since it is easier to teach a man to handle a tool, to read a dial, than it is to teach him to ask, always, for a name-brand aspirin—to want, someday, a stand-by generator.

What is that? You don't know? I used not to know, but the readers of *House Beautiful* all know, so that now I know. It is the electrical generator that stands in the basement of the suburban house-owner, shining, silent, till at last one night the lights go out, the furnace stops, the freezer's food begins to—

Ah, but it's frozen for good, the lights are on forever; the owner has switched on the stand-by generator.

But you don't see that he really needs the generator, you'd rather have seen him buy a second car? He has two. A second bathroom? He has four. When the People of the Medium doubled everything, he doubled everything; and now that he's gone twice round he will have to wait three years, or four, till both are obsolescent—but while he waits there are so many new needs that he can satisfy, so many things a man can buy. "Man wants but little here below/ Nor wants that little long," said the poet; what a lie! Man wants almost unlimited quantities of almost everything, and he wants it till the day he dies.

Sometimes in *Life* or *Look* we see a double-page photograph of some family standing on the lawn among its possessions: station-wagon, swimming-pool, power-cruiser, sports-car, tape-recorder, television sets, radios, cameras, power lawn-mower, garden tractor, lathe, barbecue-set, sporting equipment, domestic appliances—all the gleaming, grotesquely imaginative paraphernalia of its existence. It was hard to get everything on two pages, soon it will need four. It is like a dream, a child's dream before Christmas; yet if the members of the family doubt that they are awake, they have only to reach out and pinch something. The family seems pale and small, a negligible appendage, beside its possessions; only a human being would need to ask: "Which owns which?" We are fond of saying that something is not just something but "a way of life"; this too is a way of life—our way, the way.

Emerson, in his spare stony New England, a few miles from Walden, could write: "Things are in the saddle/ And ride mankind." He could say more now: that they are in the theater and studio, and entertain mankind; are in the pulpit and preach to mankind. The values of business, in a business society like our own, are reflected in every sphere: values which agree with them are reinforced, values which disagree are cancelled out or have lip service paid to them. In business what sells is good, and that's the end of it—that is what *good* means; if the world doesn't beat a path to your door, your mouse-trap wasn't better. The values of the Medium—which is both a popular business itself and the cause of popularity in other businesses—are business values: money, success, celebrity. If we are representative members of our society, the Medium's values are ours; and even if we are unrepresentative, non-conforming, our hands are—too often—subdued to the element they work in, and our unconscious expectations are all that we consciously reject. Darwin said that he always immediately wrote down evidence against a theory because otherwise, he'd noticed, he would forget it; in the same way, we keep forgetting the existence of those poor and unknown failures whom we might rebelliously love and admire.

If you're so smart why aren't you rich? is the ground-bass of our society, a grumbling and quite unanswerable criticism, since the society's non-monetary values *are* directly convertible into money. Celebrity turns into testimonials, lectures, directorships, presidencies, the capital gains of an autobiography *Told To* some professional ghost who photographs the man's life as Bachrach photographs his body. I read in the newspapers a lyric and perhaps exaggerated instance of this direct conversion of celebrity into money: his son accompanied Adlai Stevenson on a trip to Russia, took snapshots of his father, and sold them (to accompany his father's account of the trip) to *Look* for $20,000. When Liberace said that his critics' unfavorable reviews hurt him so much that he cried all the way to the bank, one had to admire the correctness and penetration of his press-agent's wit—in another age, what might not such a man have become!

Our culture is essentially periodical: we believe that all that is deserves to perish and to have something else put in its place. We speak of planned obsolescence, but it is more than planned, it is felt; is an assumption about the nature of the world. We feel that the present is better and more interesting, more real, than the past, and that the future will be better and more interesting, more real, than the present; but, consciously, we do not hold against the present its prospective obsolescence. Our standards have become to an astonishing degree the standards of what is called the world of fashion, where mere timeliness—being orange in orange's year, violet in violet's—is the value to which all other values are reducible. In our society the word *old-fashioned* is so final a condemnation that someone like Norman Vincent Peale can say about atheism or agnosticism simply that it is old-fashioned; the homely recommendation of the phrase *Give me that good old-time religion* has become, after a few decades, the conclusive rejection of the phrase *old-fashioned atheism.*

All this is, at bottom, the opposite of the world of the arts, where commercial and scientific progress do not exist; where the bone of Homer and Mozart and Donatello is there, always, under the mere blush of fashion; where the past—the remote past, even—is responsible for the way that we understand, value, and act in, the present. (When one reads an abstract expressionist's remark that Washington studios are "eighteen months behind" those of his colleagues in New York, one realizes something of the terrible power of business and fashion over those most overtly hostile to them.) An artist's work and life presuppose continuing standards, values extended over centuries or millenia, a future that is the continuation and modification of the past, not its contradiction or irrelevant replacement. He is working for the time that wants the best that he can do: the present, he hopes—but if not that, the future. If he sees that fewer and fewer people are any real audience for the serious artists of the past, he will feel that still fewer are going to be an audience for the serious artists of the present: for those who, willingly or unwillingly, sacrifice extrinsic values to intrinsic ones, immediate effectiveness to that steady attraction which, the artist hopes, true excellence will always exert.

The past's relation to the artist or man of culture is almost the opposite of its relation to the rest of our society. To him the present is no more than the last ring on the trunk, understandable and valuable only in terms of all the earlier rings. The rest of our society sees only that great last ring, the enveloping surface of the trunk; what's underneath is a disregarded, almost mythical foundation. When Northrop Frye writes that "the preoccupation of the humanities with the past is sometimes made a reproach against them by those who forget that we face the past: it may be shadowy, but it is all that is there," he is saying what for the artist or man of culture is self-evidently true. Yet for the Medium and the People of the Medium it is as self-evidently false: for them the present—or a past so recent, so quick-changing, so soon-disappearing, that it might be called the specious present—is all that is there.

In the past our culture's body of common knowledge—its frame of reference, its possibility of comprehensible allusion—changed slowly and superficially; the amount added to it or taken away from it, in any ten years, was surprisingly small. Now in any ten years a surprisingly large proportion of the whole is replaced. Most of the information people have in common is something that four or five years from now they will not even remember having known. A newspaper story remarks in astonishment that television quiz-programs "have proved that ordinary citizens can be conversant with such esoterica as jazz, opera, the Bible, Shakespeare, poetry, and fisticuffs." You may exclaim: "Esoterica! If the Bible and Shakespeare are esoterica, what is there that's common knowledge?" The answer, I suppose, is that Elfrida von Nordroff and Teddy Nadler—the ordinary citizens on the quiz-programs—are common knowledge; though not for long. Songs disappear in two or three months, celebrities in two or three years; most of the Medium is little felt and soon forgotten. Nothing is as dead as day-before-yesterday's newspaper, the next-to-the-last number on the roulette wheel; but most of the knowledge people have in common and lose in common is knowledge of such newspapers, such numbers. Yet the novelist or poet or dramatist, when he moves a great audience, depends upon the deep feelings, the living knowledge, that the people of that audience share; if so much has become contingent, superficial, ephemeral, it is disastrous for him.

New products and fashions replace the old, and the fact that they replace them is proof enough of their superiority. Similarly, the Medium does not need to show that the subjects which fill it are interesting or timely or important; the fact that they are its subjects makes them so. If *Time*, *Life*, and the television shows are full of Tom Fool this month, he's no fool. And when he has been gone from them a while, we do not think him a fool—we do not think of him at all. He no longer exists, in the fullest sense of the word *exist:* to be is to be perceived, to be a part of the Medium of our perception. Our celebrities are not kings, romantic in exile, but Representatives who, defeated, are forgotten; they had, always, only the qualities that we delegated to them.

After driving for four or five minutes along the road outside my door, I come to a row of one-room shacks about the size of kitchens, made out of used boards, metal signs, old tin roofs. To the people who live in them an electric dishwasher of one's own is as much a fantasy as an ocean liner of one's own. But since the Medium (and those whose thought is molded by it) does not perceive them, these people are themselves a fantasy. No matter how many millions of such exceptions to the general rule there are, they do not really exist, but have a kind of anomalous, statistical subsistence; our moral and imaginative view of the world is no more affected by them than by the occupants of some home for the mentally deficient a little farther along the road. If some night one of these out-moded, economically deficient ghosts should scratch at my window, I could say only: "Come back twenty or thirty years ago." And if I myself, as an old-fashioned, one-room poet, a friend of "quiet culture," a "meek lover of the good," should go out some night to scratch at another window, shouldn't I hear someone's indifferent or regretful: "Come back a century or two ago"?

When those whose existence the Medium recognizes ring the chimes of the writer's doorbell, fall through his letter-slot, float out onto his television-screen, what is he to say to them? A man's unsuccessful struggle to get his family food is material for a work of art—for tragedy, almost; his unsuccessful struggle to get his family a stand-by generator is material for what? Comedy? Farce? Comedy on such a scale, at such a level, that our society and its standards seem, almost, farce? And yet it is the People of the Medium—those who struggle for and get, or struggle for and don't get, the generator—whom our society finds representative: they are there, there primarily, there to be treated first of all. How shall the artist treat them? And the Medium itself—an end of life and a means of life, something essential to people's understanding and valuing of their existence, something many of their waking hours are spent listening to or looking at—how is *it* to be treated as subject-matter for art? The artist cannot merely reproduce it; should he satirize or parody it? But by the time the artist's work reaches its audience, the portion of the Medium which it satirized will already have been forgotten; and parody is impossible, often, when so much of the Medium is already an unintentional parody. (Our age might be defined as the age in which real parody became impossible, since any parody had already been duplicated, or parodied, in earnest.) Yet the Medium, by now, is an essential part of its watchers. How can you explain those whom Mohammedans call the People of the Book in any terms that omit the Book? We are people of the television-set, the magazine, the radio, and are inexplicable in any terms that omit them.

Oscar Wilde said that Nature imitates Art, that before Whistler painted them there were no fogs along the Thames. If his statement were not false, it would not be witty. But to say that Nature imitates Art, when the Nature is human nature and the Art that of television, radio, motion-pictures, magazines, is literally true. The Medium shows its People what life is, what people are, and its People believe it: expect people to be that, try themselves to be that. Seeing is believing; and if what you see in *Life* is different from what you see in life, which of the two are you to believe? For many people it is what you see in *Life* (and in the movies, over television, on the radio) that is real life; and everyday existence, mere local or personal variation, is not real in the same sense.

The Medium mediates between us and raw reality, and the mediation more and more replaces reality for us. Many radio-stations have a newsbroadcast every hour, and many people like and need to hear it. In many houses either the television set or the radio is turned on during most of the hours the family is awake. It is as if they longed to be established in reality, to be reminded continually of the "real," "objective" world—the created world of the Medium—rather than to be left at the mercy of actuality, of the helpless contingency of the world in which the radio-receiver or television set is sitting. And surely we can sympathize: which of us hasn't found a similar refuge in the "real," created world of Cézanne or Goethe or Verdi? Yet Dostoievsky's world is too different from Wordsworth's, Piero della Francesca's from Goya's, Bach's from Wolf's, for us to be able to substitute one homogeneous mediated reality for everyday reality in the belief that it *is* everyday reality. For many watchers, listeners, readers, the world of events and celebrities and performers—the Great World—has become the world of primary reality: how many times they have sighed at the colorless unreality of their own lives and families, and sighed for the bright reality of, say, Elizabeth Taylor's. The watchers call the celebrities by their first names, approve or disapprove of "who they're dating," handle them with a mixture of love, identification, envy, and contempt. But however they handle them, they *handle* them: the Medium has given everyone so terrible a familiarity with everyone that it takes great magnanimity of spirit not to be affected by it. These celebrities are not heroes to us, their valets.

Better to have these real ones play themselves, and not sacrifice too much of their reality to art; better to have the watcher play himself, and not lose too much of himself in art. Usually the watcher is halfway between two worlds, paying full attention to neither: half distracted from, half distracted by, this distraction; and able for the moment not to be too greatly affected, have too great demands made upon him, by either world. For in the Medium, which we escape to from work, nothing is ever *work*, makes intellectual or emotional or imaginative demands which we might find it difficult to satisfy. Here in the half-world everything is homogeneous—is, as much as possible, the same as everything else: each familiar novelty, novel familiarity has the same treatment on top and the same attitude and conclusion at bottom; only the middle, the particular subject of the particular program or article, is different. If it *is* different: everyone is given the same automatic "human interest" treatment, so that it is hard for us to remember, unnecessary for us to remember, which particular celebrity we're reading about this time—often it's the same one, we've just moved on to a different magazine.

Francesco Caraccioli said that the English have a hundred religions and one sauce; so do we; and we are so accustomed to this sauce or dye or style of presentation, the aesthetic equivalent of Standard Brands, that a very simple thing can seem obscure or perverse without it. And, too, we find it hard to have to shift from one genre to another, to vary our attitudes and expectations, to use our unexercised imaginations. Poetry disappeared long ago, even for most intellectuals; each year fiction is a little less important. Our age is the age of articles: we buy articles in stores, read articles in magazines, exist among the interstices of articles: of columns, interviews, photographic essays, documentaries; of facts condensed into headlines or expanded into non-fiction best-sellers; of real facts about real people.

Art lies to us to tell us the (sometimes disquieting) truth. The Medium tells us truths, facts, in order to make us believe some reassuring or entertaining lie or half-truth. These actually existing celebrities, of universally admitted importance, about whom we are told directly authoritative facts —how can fictional characters compete with these? These *are* our fictional characters, our Lears and Clytemnestras. (This is ironically appropriate, since many of their doings and sayings are fictional, made up by public relations officers, columnists, agents, or other affable familiar ghosts.) And the Medium gives us such facts, such tape-recordings, such clinical reports not only about the great but also about (representative samples of) the small. When we have been shown so much about so many—*can* be shown, we feel, anything about anybody—does fiction seem so essential as it once seemed? Shakespeare or Tolstoy can show us all about someone, but so can *Life;* and when *Life* does, it's someone real.

The Medium is half life and half art, and competes with both life and art. It spoils its audience for both; spoils both for its audience. For the People of the Medium life isn't sufficiently a matter of success and glamor and celebrity, isn't entertaining enough, distracting enough, *mediated* enough; and art is too difficult or individual or novel, too much a matter of tradition and the past, too much a matter of special attitudes and aptitudes—its mediation sometimes is queer or excessive, and sometimes is not even recognizable as mediation. The Medium's mixture of rhetoric and reality, in which people are given what they know they want to be given in the form in which they know they want to be given it, is something more efficient and irresistible than any real art. If a man has all his life been fed a combination of marzipan and ethyl alcohol—if eating, to him, is a matter of being knocked unconscious by an ice cream soda—can he, by taking thought, come to prefer a diet of bread and wine, apples and well-water? Will a man who has spent his life watching gladiatorial games come to prefer listening to chamber music? And those who produce the bread and the wine and the quartets for him—won't they be tempted either to give up producing them, or else to produce a bread that's half sugar and half alcohol, a quartet that ends with the cellist at the violist's bleeding throat?

Any outsider who has worked for the Medium will have observed that the one thing which seems to its managers most unnatural is for someone to do something naturally, to speak or write as an individual speaking or writing to other individuals, and not as a sub-contractor supplying a standardized product to the Medium. It is as if producers and editors and supervisors—middle men—were particles forming a screen between maker and public, one which will let through only particles of their own size and weight (or as they say, the public's). As you look into their strained puréed faces, their big horn-rimmed eyes, you despair of Creation itself, which seems for the instant made in their own owl-eyed image. There are so many extrinsic considerations involved in the presentation of his work, the maker finds, that by the time it is presented almost any intrinsic consideration has come to seem secondary. No wonder that the professional who writes the ordinary commercial success—the ordinary script, scenario, or best seller—resembles imaginative writers less than he resembles editors, producers, executives. The supplier has come to resemble those he supplies, and what he supplies them resembles both. With an artist you never know what you will get; with him you know what you will get. He is a reliable source for a standard product. He is almost exactly the opposite of the imaginative artist: instead of stubbornly or helplessly sticking to what he sees and feels—to what is right for him, true to his reality, regardless of what the others think and want—he gives the others what they think and want, regardless of what he himself sees and feels.

The Medium represents, to the artist, all that he has learned not to do: its sure-fire stereotypes seem to him what any true art, true spirit, has had to struggle past on its way to the truth. The artist sees the values and textures of this art-substitute replacing those of his art, so far as most of society is concerned; conditioning the expectations of what audience his art has kept. Mass culture either corrupts or isolates the writer. His old feeling of oneness—of speaking naturally to an audience with essentially similar standards—is gone; and writers no longer have much of the consolatory feeling that took its place, the feeling of writing for the happy few, the kindred spirits whose standards are those of the future. (Today they feel: the future, should there be one, will be worse.) True works of art are more and more produced away from or in opposition to society. And yet the artist needs society as much as society needs him: as our cultural enclaves get smaller and drier, more hysterical or academic, one mourns for the artists inside and the public outside. An incomparable historian of mass culture, Ernest van den Haag, has expressed this with laconic force: "The artist who, by refusing to work for the mass market, becomes marginal, cannot create what he might have created had there been no mass market. One may prefer a monologue to addressing a mass meeting. But it is still not a conversation."

Even if the rebellious artist's rebellion is whole-hearted, it can never be whole-stomach'd, whole-unconscious'd. Part of him wants to be like his kind, is like his kind; longs to be loved and admired and successful. Our society—and the artist, in so far as he is truly a part of it—has no place set aside for the different and poor and obscure, the fools for Christ's sake: they all go willy-nilly into Limbo. The artist is tempted, consciously, to give his society what it wants—or if he won't or can't, to give it nothing at all; is tempted, unconsciously, to give it superficially independent or contradictory works which are at heart works of the Medium. But it is hard for him to go on serving both God and Mammon when God is so really ill-, Mammon so really well-organized.

"Shakespeare wrote for the Medium of his day; if Shakespeare were alive now he'd be writing *My Fair Lady;* isn't *My Fair Lady*, then, our *Hamlet?* shouldn't you be writing *Hamlet* instead of sitting there worrying about your superego? I need my *Hamlet!*" So society speaks to the artist, reasons with the artist; and after he has written it its *Hamlet* it is satisfied, and tries to make sure that he will never do it again. There are many more urgent needs that it wants him to satisfy: to lecture to it; to be interviewed; to appear on television programs; to give testimonials; to attend book luncheons; to make trips abroad for the State Department; to judge books for Book Clubs; to read for publishers, judge for publishers, be a publisher for publishers; to edit magazines; to teach writing at colleges or conferences; to write scenarios or scripts or articles—articles about his home town for *Holiday*, about cats or clothes or Christmas for *Vogue*, about "How I Wrote *Hamlet*" for anything; to—

But why go on? I once heard a composer, lecturing, say to a poet, lecturing: "They'll pay us to do *anything*, so long as it isn't writing music or writing poems." I knew the reply that as a member of my society I should have made: "As long as they pay you, what do you care?" But I didn't make it: it was plain that they cared . . . But how many more learn not to care, to love what they once endured! It is a whole so comprehensive that any alternative seems impossible, any opposition irrelevant; in the end a man says in a small voice: "I accept the Medium." The Enemy of the People winds up as the People—but where there is no enemy, the people perish.

The climate of our culture is changing. Under these new rains, new suns, small things grow great, and what was great grows small; whole species disappear and are replaced. The American present is very different from the American past: so different that our awareness of the extent of the changes has been repressed, and we regard as ordinary what is extraordinary—ominous perhaps—both for us and for the rest of the world. The American present is many other peoples' future: our cultural and economic example is to much of the world mesmeric, and it is only its weakness and poverty that prevent it from hurrying with us into the Roman future. But at this moment of our power and success, our thought and art are full of a troubled sadness, of the conviction of our own decline. When the President of Yale University writes that "the ideal of the good life has faded from the educational process, leaving only miscellaneous prospects of jobs and joyless hedonism," are we likely to find it unfaded among our entertainers and executives? Is the influence of what I have called the Medium likely to lead us to any good life? to make us love and try to attain any real excellence, beauty, magnanimity? or to make us understand these as obligatory but transparent rationalizations behind which the realities of money and power are waiting?

The tourist Matthew Arnold once spoke about our green culture in terms that have an altered relevance—but are not yet irrelevant—to our ripe one. He said: "What really dissatisfies in American civilization is the want of the *interesting*, a want due chiefly to the want of those two great elements of the interesting, which are elevation and beauty." This use of *interesting*—and, perhaps, this tone of a curator pointing out what is plain and culpable—shows how far along in the decline of West Arnold came: it is only in the latter days that we ask to be interested. He had found the word, he tells us, in Carlyle. Carlyle is writing to a friend to persuade him not to emigrate to the United States; he asks: "Could you banish yourself from all that is interesting to your mind, forget the history, the glorious institutions, the noble principles of old Scotland—that you might eat a better dinner, perhaps?" We smile, and feel like reminding Carlyle of the history, the glorious institutions, the noble principles of new America—of that New World which is, after all, the heir of the Old.

And yet . . . Can we smile as comfortably, to-day, as we could have smiled yesterday? Nor could we listen as unconcernedly, if on taking leave of us some other tourist should conclude, with the penetration and obtuseness of his kind:

"I remember reading somewhere: that which you inherit from your fathers you must earn in order to possess. I have been so much impressed with your power and your possessions that I have neglected, perhaps, your principles. The elevation or beauty of your spirit did not equal, always, that of your mountains and skyscrapers: it seems to me that your society provides you with 'all that is interesting to the mind' only exceptionally, at odd hours, in little reservations like those of your Indians. But as for your dinners, I've never seen anything like them: your daily bread comes *flambé*. And yet—wouldn't you say—the more dinners a man eats, the more comforts he possesses, the hungrier and more uncomfortable some part of him becomes: inside every fat man there is a man who is starving. Part of you is being starved to death, and the rest of you is being stuffed to death. But this will change: no one goes on being stuffed to death or starved to death forever.

"This is a gloomy, an equivocal conclusion? Oh yes, I come from an older culture, where things are accustomed to coming to such conclusions; where there is no last-paragraph fairy to bring one, always, a happy ending—or that happiest of all endings, no ending at all. And have I no advice to give you as I go? None. You are too successful to need advice, or to be able to take it if it were offered; but if ever you should fail, it is there waiting for you, the advice or consolation of all the other failures."

Poets, Critics, and Readers

PEOPLE often ask me: "Is there any poet who makes his living writing poetry?" and I have to say: "No." The public has an unusual relationship to the poet: it doesn't even know that he is there. Our public is a rich and generous one; if it knew that the poet was there, it would pay him for being there. As it is, poets make their living in many ways: by being obstetricians, like William Carlos Williams; or directors of Faber and Faber, like T. S. Eliot; or vice-presidents of the Hartford Accident and Indemnity Company, like Wallace Stevens. But most poets, nowadays, make their living by teaching. Kepler said, "God gives every animal a way to make its living, and He has given the astronomer astrology"; and now, after so many centuries, He has given us poets students. But what He gives with one hand He takes away with the other: He has taken away our readers.

Yet the poet can't help looking at what he has left, his students, with gratitude. His job may be an impossible one—there are three impossible tasks, said Freud: to teach, to govern, and to cure —but what is there so grateful as impossibility? and what is there better to teach, more nearly impossible to teach, than poems and stories? As Lord Macaulay says: "For how can man live better/ Than facing fearful odds/ For the poems of his fathers—"

I seem to have remembered it a little wrong, but it's a natural error. And, today, when we get people to read poems—to read very much of anything —naturally and joyfully, to read it not as an unnatural rightness but as a natural error: what people always have done, always will do—we do it against fearful odds. I can't imagine a better way for the poet to make his living. I certainly can't imagine his making his living by writing poems—I'm not *that* imaginative. I'm used to things as they are.

But there is a passage in Wordsworth that I read, always, with a rueful smile. He is answering the question, *Why write in verse?* He gives several reasons. His final reason, he writes, "is all that is *necessary* to say upon this subject." Here it is, all that it is *necessary* to say upon this subject: "Few persons will deny, that of two descriptions, either of passions, manners, or characters, each of them equally well executed, the one in prose and the other in verse, the verse will be read a hundred times where the prose is read once."

One sees sometimes, carved on geology buildings: *O Earth, what changes thou hast seen!* When a poet finishes reading this passage from Wordsworth, he thinks in miserable awe: *O Earth, what changes thou hast seen!* Only a hundred and fifty years ago *this* is what people were like. Nowadays, of course, the prose will be read a thousand times where the verse is read once. And this seems to everybody only natural; the situation Wordsworth describes seems unnatural, improbable, almost impossible. What Douglas Bush writes is true: we live in "a time in which most people assume that, as an eminent social scientist once said to me, 'Poetry is on the way out.' " To most of us verse, any verse, is so uncongenial, so exhaustively artificial, that I have often thought that a man could make his fortune by entirely eliminating from our culture verse of any kind: in the end there would *be* no more poems, only prose translations of them. This man could begin by publishing his Revised Standard Version of *Mother Goose:* without rhyme, meter, or other harmful adulterants; with no word of anything but honest American prose, prose that cats and dogs can read.

A friend of mine once took a famous Italian scholar on a tour of New Haven. She specialized in objects of art and virtue—samplers, figureheads, paintings of women under willows, statues of General Washington—but no matter what she showed him, the man would only wave his hand in the air and exclaim: *Ridickalus!* And shouldn't we feel so about things like *Mother Goose?*

Early to bed and early to rise
Makes a man healthy, wealthy, and wise.

Ridickalus! Why say it like a rocking horse? why make it jingle so? and *wise*—who wants to be wise?

> *Which sibling is the well-adjusted sibling?*
> *The one that gets its sleep.*

That is the way the modern *Mother Goose* will put it. I don't expect the modern *Mother Goose* to be especially popular with little children, who have not yet learned not to like poetry; but it is the parents who buy the book.

Isn't writing verse a dying art, anyway, like blacksmithing or buggymaking? Well, not exactly: poets are making as many buggies as ever—good buggies, fine buggies—they just can't get anybody much to ride in them. As for blacksmithing: I read the other day that there are twice as many blacksmith shops in the United States as there are bookstores. Something has gone wrong with that comparison too. No, I'm doing what poets do, complaining; and if I exaggerate a little when I complain, why, that's only human—surely you want me to exaggerate a little, in my misery. Goethe says, when he is talking about slum children: "No person ever looks miserable who feels that he has the right to make a demand on you." This right is not anything that anyone can confer upon himself; it is the public, society, all of us, that confer this right. If the poet looks miserable, it is because we have made him feel that he no longer has the right to make a demand on us. It is no longer a question of what he wants, or of what he ought to be given—he takes what he gets, and complains about getting it, and he hears the echo of his complaint, and then the silence settles around him, a little darker, a little deeper.

What does he want? To be read. Read by whom? critics? men wise enough to tell him, when they have read the poem, what it is and ought to be, what its readers feel and ought to feel? Well, no. A writer cannot learn about his readers from his critics: they are different races. The critic, unless he is one in a thousand, reads to criticize; the reader reads to read.

Freud talks of the "free-floating" or "evenly-hovering" attention with which the analyst must listen to the patient. Concentration, note-taking, listening with a set—a set of pigeonholes—makes it difficult or impossible for the analyst's unconscious to respond to the patient's; takes away from the analyst the possibility of learning from the patient what the analyst doesn't already know; takes away from him all those random guesses or intuitions or inspirations which come out of nowhere—and come, too, out of the truth of the patient's being. But this is quite as true of critics and the poems that are *their* patients: when one reads as a linguist, a scholar, a New or Old or High or Low critic, when one reads the poem *as a means to an end*, one is no longer a pure reader but an applied one. The true reader "listens like a three years' child:/ The Mariner hath his will." Later on he may write like a sixty-three-year-old sage, but he knows that in the beginning, unless ye be converted, and become as little children, ye shall not enter into the kingdom of art. Hofmannsthal says, with awful finality: "The world has lost its innocence, and without innocence no one creates or enjoys a work of art"; but elsewhere he says more hopefully, with entire and not with partial truth, that each of us lives in an innocence of his own which he never entirely loses.

Is there a public for poetry that is still, in this sense of the word, innocent? Of course, there are several publics for poetry—small, benighted, eccentric publics—just as there are publics for postage stamps and cobblers' benches; but this is such a disastrous change from the days of *Childe Harold* and *In Memoriam* and *Hiawatha*, when the public for poetry was, simply, the reading public, that you can see why poets feel the miserable astonishment that they feel. The better-known poets feel it more than the lesser-known, who—poor things—lie under the table grateful for crumbs, pats, kicks, anything at all that will let them be sure they really exist, and are not just a dream someone has stopped dreaming. A poet like Auden says that nobody reads him except poets and young men in cafeterias —his description of the young men is too repellent for me to repeat it to you.

Literally, Auden is wrong: we read Auden, this is no cafeteria; but, figuratively, Auden is right—the poet's public's gone. Frederick the Great translated Voltaire, and trembled as the poet read the translation; Elizabeth—Elizabeth the First—and Henry the Eighth and Richard the Lion-hearted wrote good poems and read better; and I cannot resist quoting to you three or four sentences from Frans Bengtsson's novel *The Long Ships*, to show you what things were like at the court of Harald Bluetooth, King of Denmark in the year 1000. A man gets up from a banquet table: "His name was Björn Asbrandsson, and he was a famous warrior, besides being a great poet to boot. . . . Although he was somewhat drunk, he managed to improvise some highly skilful verses in King Harald's honor in a meter known as *töglag*. This was the latest and most difficult verse-form that the Icelandic poets had invented, and indeed the poem was so artfully contrived that little could be understood of its content. Everybody, however, listened with an appearance of understanding, for any man who could not understand poetry would be regarded as a poor specimen of a warrior; and King Harald praised the poet and gave him a gold ring."

Auden is a descendant of just such poets as this one; but if Auden, when he next visits the University of your state, makes up an incomprehensible poem, in a difficult new meter, in honor of the President of the University, will all its football players pretend they understand the poem, so as not to be thought poor specimens of football players? and will the President give Auden a gold ring?

In the days when his readers couldn't read, the

poet judged his public by his public: the gold ring or the scowl the king gave him was as concrete as the labored, triumphant faces of his hearers. But nowadays King Harald and his warriors are represented by a reviewer, next year, in the New York *Times;* a critic, nine years later, in the *Sewanee Review.* "Ah, better to sing my songs to a wolf pack on the Seeonee than to a professor on the *Sewanee!*" the poet blurts, baring his teeth; but then—what choice has he?—he lets the Reality Principle do its worst, and projects or extends or extrapolates a critic or two, a dozen reviewers, into the Public; into Posterity. Critics, alas! are the medium through which the poet darkly senses his public. Nor is it altogether different for the public: Harald and his vikings, lonely in their split-levels, do not even remember the days when, as they listened, they could look into one another's faces and know without looking what they would find there. Now they too look into the *Times;* wish that they could replace that scowl with a gold ring, that gold ring with a scowl; reconstruct from the exclamations on dust jackets, quotations in advertisements, the fierce smiles on the faces of the warriors.

So if we are to talk about the poet and his poems and his public, what each is to the others, we must spend much of our time—too much of our time—talking about his critics. Criticism is necessary, I suppose; I know. Yet criticism, to the poet, is no necessity, but a luxury he can ill afford. Conrad cried to his wife: "I don't want criticism, I want praise!"

And it is praise, blame, tears, laughter, that writers want; when Columbus comes home he needs to be cheered for finding a new way to India, not interned while the officials argue about whether it is Asia, Africa, or Antarctica that he has discovered. Really, of course, it's America—and if they agreed about it this would be helpful to Columbus; he could say to himself, in awe: "So it was America I discovered!" But how seldom the critics do agree! A gray writer seems black to his white critics, white to his black critics: the same poem will seem incomprehensible modernistic nonsense to Robert Hillyer, and a sober, old-fashioned, versified essay to the critics of some little magazine of advanced tastes. Ordinary human feeling, the most natural tenderness, will seem to many critics and readers rank sentimentality, just as a kind of nauseated brutality (in which the writer's main response to the world is simply to vomit) will seem to many critics and readers the inescapable truth. We live in a time in which Hofmannsthal's "Good taste is the ability continuously to counteract exaggeration" will seem to most readers as false as it seems tame. "Each epoch has its own sentimentality," Hofmannsthal goes on, "its specific way of overemphasizing strata of emotion. The sentimentality of the present is egotistic and unloving; it exaggerates not the feeling of love but that of the self."

Everyone speaks of the "negative capability" of the artist, of his ability to lose what self he has in the many selves, the great self of the world. Such a quality is, surely, the first that a critic should have; yet who speaks of the negative capability of the critic? how often are we able to observe it? The commonest response to the self of a work of art is the critic's assertion that he too has a self. What he writes proves it. I once saw, in an essay by a psychoanalyst, the phrase *the artist and his competitor, the critic*. Where got he that truth? Out of an analysand's mouth? I do not know; but that it is an important and neglected truth I do know. All mediators become competitors: the exceptions to this rule redeem their kind.

Critics disagree about almost every quality of a writer's work; and when some agree about a quality, they disagree about whether it is to be praised or blamed, nurtured or rooted out. After enough criticism the writer is covered with lipstick and bruises, and the two are surprisingly evenly distributed. There is *nothing* so plain about a writer's books, to some critics, that its opposite isn't plain to others. Kafka is original? Not at all, according to Edmund Wilson. A fine critic of poetry, Ezra Pound, writes: "In [the writer So-and-So] you have an embroidery of language, a talk *about* the matter, not presentation; you have grace, richness of language, etc., as much as you like, but you have nothing that isn't replaceable by something else, no ornament that wouldn't have done just as well in some other connection, or that for which some other figure of rhetoric or fancy couldn't have served, or which couldn't have been distilled from

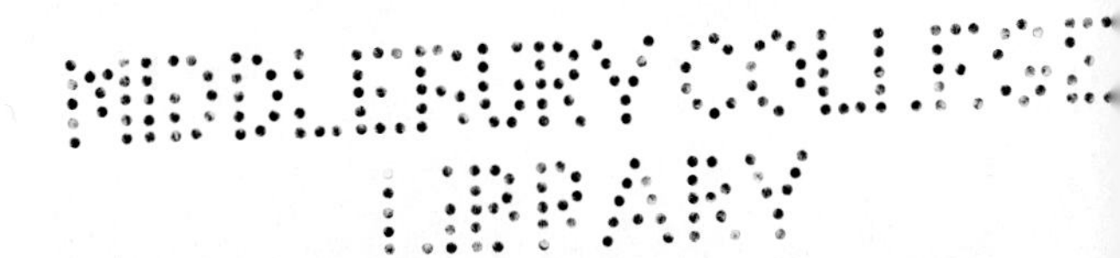

literary antecedents." About whom is Pound speaking? About Shakespeare. Anyone who has read at all widely has come across thousands of such judgments, and it is easy for him to sympathize with the artist when the artist murmurs: "We wish to learn from our critics, but it is hard for us even to recover from them. A fool's reproach has an edge like a razor, and his brother's praise is small consolation. Critics are like bees: one sting lasts longer than a dozen jars of honey."

The best thing ever said about criticism—I am not, now, speaking as a critic—was said, as is often the case, by Goethe: "Against criticism we can neither protect nor defend ourselves; we must act in despite of it, and gradually it resigns itself to this." The great Goethe suffered just as we little creatures do, and he spoke about it, as we don't, in imperishable sentences: "All great excellence in life or art, at its first recognition, brings with it a certain pain arising from the strongly felt inferiority of the spectator; only at a later period, when we take it into our own culture, and appropriate as much of it as our capacities allow, do we learn to love and esteem it. Mediocrity, on the other hand, may often give us unqualified pleasure; it does not disturb our self-satisfaction, but rather encourages us with the thought that we are as good as another. . . . Properly speaking, we learn only from those books we cannot judge. The author of a book that I am competent to criticize would have to learn from me."

Goethe says over and over: "Nothing is more ter-

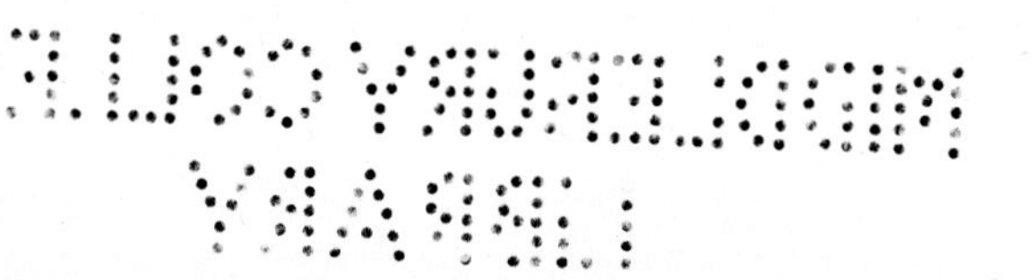

rible than ignorance in action. . . . It is a terrible thing when fools thrive at the expense of a superior man." You and I will agree—and then we will have to decide whether we're being thriven at the expense of, or thriving. Goethe says in firm doggerel: "However clear and simple be it/ Finder and doer alone may see it." No, Goethe didn't have too much use for critics, since he thought that critics weren't of too much use.

And why am I quoting all this to you? have critics hurt me so that I want to pull down the temple upon their heads, even if I too perish in the ruins?—for I too am a critic. No, it's not that; critics have done their best for me, and their best has been, perhaps, only too good; when I myself criticize, I am willing for you to believe what I say; but I am trying to explain why it is that critics are of so little use to writers, why it is that they are such a poor guide to the opinions of the next age—and I am explaining in an age which has an unprecedented respect for, trust in, criticism.

All of us have read pieces of criticism—many pieces of criticism—which seem worthy both of delighted respect and cautious trust. All of us have read criticism in which the critic takes it for granted that what he writes about comes first, and what he writes comes second—takes it for granted that he is writing as a reader to other readers, to be of use to them; criticism in which the critic works, as far as he is able, in the spirit of Wordsworth's "I have endeavored to look steadily at my subject." All of us have some favorite, exceptional critic who might say, with substantial truth, that he has not set up rigid standards to which a true work of art must conform, but that he has tried instead to let the many true works of art—his experience of them—set up the general expectations to which his criticism of art conforms; that he has tried never to see a work of art as mere raw material for criticism, data for generalization; that he has tried never to forget the difference between creating a work of art and criticising a work of art; and that he has tried, always, to remember what Proust meant when he said, about writers like Stendhal, Balzac, Hugo, Flaubert, the great creators called "romantics": "The classics have no better commentators than the 'romantics.' The romantics are the only people who really know how to read the classics, because they read them as they were written, that is to say, 'romantically' and because if one would read a poet or a prose writer properly one must be, not a scholar, but a poet or a prose writer." It might be put a little differently: if one would read a poet or a prose writer properly one must be, not a scholar or a poet or a prose writer, but a reader: someone who reads books as they were written, that is to say, "romantically." Proust's grandmother was not a poet or a prose writer, but she read Madame de Sévigné properly. To be, as she was, a reader, is a lofty and no longer common fate.

The best poetry critic of our time, T. S. Eliot,

has said about his criticism: "I see that I wrote best about poets whose work had influenced my own, and with whose poetry I had become thoroughly familiar, long before I desired to write about them, or had found the occasion to do so. . . . The best of my literary criticism . . . is a by-product of my private poetry-workshop." But perhaps something of this sort is always true: perhaps true criticism is something, like sincerity or magnanimity, that cannot be aimed at, attained, directly; that must always be, in some sense, a by-product, whether of writing or reading, of a private poetry-workshop or a private reading-room.

We all realize that writers are inspired, but helpless and fallible beings, who know not what they write; readers, we know from personal experience, are less inspired but no less helpless and fallible beings, who half the time don't know what they're reading. Now, a critic is half writer, half reader: just as the vices of men and horses met in centaurs, the weaknesses of readers and writers meet in critics. A good critic—we cannot help seeing, when we look back at any other age—is a much rarer thing than a good poet or a good novelist. Unless you are one critic in a hundred thousand, the future will quote you only as an example of the normal error of the past, what everybody was foolish enough to believe then. Critics are discarded like calendars; yet, for their year, with what trust the world regards them!

Art is long, and critics are the insects of a day.

But while he survives, it is the work of art he criticizes which is the critic's muse, or daemon, or guardian angel: it is a delight to the critic to think that sometimes, in moments of particular good fortune, some poem by Rilke or Yeats or Wordsworth has hovered above him, whispering what to say about it in his ear. And, in the moments of rash ambition which can come even to such humble—rightly humble—things as critics, the critic can imagine some reader, in the midst of his pleasure at a poem or story the critic has guided him to, being willing to think of some paragraph of the critic's work in terms of a sentence of Goethe's: "There is a sensitive empiricism that ultimately identifies itself with the object and thereby becomes genuine theory."

In other moments the critic can imagine the reader's thinking of him in terms of a paragraph that Proust once wrote. That miraculous writer and great critic, distressed at someone's having referred to Sainte-Beuve as one of the "great guides," exclaimed: "Surely no one ever failed so completely as did he in performing the functions of a guide? The greater part of his *Lundis* are devoted to fourth-rate writers, and whenever, by chance, he does bring himself to speak of somebody really important, of Flaubert, for instance, or Baudelaire, he immediately atones for what grudging praise he may have accorded him by letting it be understood that he writes as he does about them simply because he wants to please men who are his personal friends. . . . As to Stendhal, the novelist, the Stendhal of 'La Chartreuse,' our 'guide' laughs out of court the idea that such a person ever existed, and merely sees in all the talk about him the disastrous effects of an attempt (foredoomed to failure) to foist Stendhal on the public as a novelist. . . . It would be fun, had I not less important things to do, to 'brush in' (as Monsieur Cuvillier Fleury would have said), in the manner of Sainte-Beuve, a 'picture of French literature in the nineteenth century,' in such a way that not a single great name would appear and men would be promoted to the position of outstanding authors whose books today have been completely forgotten."

A portion of any critic, as he reads these sentences, turns white; and if another portion whispers, "Ah, but *you* needn't be afraid; certainly *you're* not as bad a critic as Sainte-Beuve," it is not a sentence to bring the color back into his cheeks, unless he blushes easily.

Wordsworth said, as Proust said after him, that "every writer, in so far as he is great and at the same time *original*, has the task of creating the taste by which he is to be enjoyed: so is it, so will it continue to be." But *taste*, he goes on to say, is a vicious and deluding word. (And surely he is right; surely we should use, instead, a phrase like *imaginative judgment.*) Using such a word as *taste* helps to make us believe that there is some passive faculty that responds to the new work of art, registering the work's success or failure; but actually the new work must call forth in us an active power analogous to that which created it—the reader "cannot proceed in quiescence, he cannot be carried like a dead weight," he must "exert himself" to feel, to sympathize, and to understand. "*There*," as Wordsworth says, "lies the true difficulty." He is right: *there* lies the difficulty for us, whether we are critics or readers; so is it, so will it continue to be.

You may say, "Of course this is true of great and original talents, but how does it apply to the trivial, immature, and eccentric writers with whom our age, like any other, is infested?" It applies only in this way: some of these trivial, immature, and eccentric writers *are* our great and original talents. The readers of Wordsworth's age said, "Of course what he says is true of great and original talents, but it is absurd when applied to a trivial and eccentric creature like Wordsworth"; and the critics of Wordsworth's age, applying the standards of the age more clearly, forcibly, and self-consciously, could condemn him with a more drastic severity. The readers read to read, the critics read to judge—both were wrong, but the critics were more impressively and rigorously and disastrously wrong, since they confirmed most readers in their dislike of Wordsworth and scared most of the others out of their liking.

We all see that the writer cannot afford to listen to critics when they are wrong—though how is he, how are we, to know when they are wrong? Can he afford to listen to them when they are right?—though how is he, how are we, to know when they are right? and right for this age or right for the next?* The writer cannot afford to question his own essential nature; must have, as Marianne Moore says, "the courage of his peculiarities." But often it is this very nature, these very peculiarities—originality always seems peculiarity, to begin with—that critics condemn. There must be about the writer a certain spontaneity or naïveté or somnambulistic rightness: he must, in some sense, move unquestioning in the midst of his world—at his question all will disappear.

And if it is slighter things, alterable things which the critics condemn, should the poet give in, alter them, and win his critics' surprised approval? "No," says Wordsworth, "where the understanding of an author is not convinced, or his feelings altered, this cannot be done without great injury to himself: for his own feelings are his stay and support, and, if he set them aside in this one instance, he may be induced to repeat this act till his mind shall lose all confidence in itself, and become utterly debilitated. To this it may be added, that the critic ought never to forget that he is himself exposed to the same errors as the Poet." Let me repeat

* "When the great innovation appears, it will almost certainly be in a muddled, incomplete, and confusing form. To the discoverer himself it will be only half-understood; to everybody else it will be a mystery. For any speculation which does not at first glance look crazy, there is no hope."

F. L. Dyson, *Innovation in Physics*

this: we ought never to forget that the critic is himself exposed to the same errors as the poet. We all know this—yet, in a deeper sense, we don't know it. We all realize that the poet's beliefs are, first of all, *his:* our books show how his epoch, his childhood, his mistresses, and his unconscious produced the beliefs; we know, now, the "real" reasons for his believing what he believed. Why do we not realize what is equally true (and equally false)?—that the critic's beliefs are, first of all, *his;* that we can write books showing how his epoch, his childhood, his mistresses, and his unconscious produced the beliefs; that we can know, now, the "real" reasons for his believing what he believed. The work of criticism is rooted in the unconscious of the critic just as the poem is rooted in the unconscious of the poet. I have had the pleasure and advantage of knowing many poets, many critics, and I have not found one less deeply neurotic than the other.

When the critic is also an artist—a T. S. Eliot—we find it easier to remember all this, and to distrust him; but when the critic is an Irving Babbitt—that is to say, a man who, tenanted by all nine of the muses, still couldn't create a couplet—we tend to think of his beliefs as somehow more objective. "Surely," we feel, "a man with so little imagination couldn't be making up something—couldn't be *inspired.*" We are wrong. Criticism is the poetry of prosaic natures (and even, in our time, of some poetic ones); there is a divinity that inspires the most sheeplike of scholars, the most tabular of critics, so that the man too dull to understand *Evangeline* still can be possessed by some theory about *Evangeline*, a theory as just to his own being as it is unjust to *Evangeline's*. The man is entitled to his inspiration; and yet . . . if only he would leave out *Evangeline!* If only he could secede from Literature, and set up some metaliterary kingdom of his own!

The poet *needs* to be deluded about his poems—for who can be sure that it is delusion? In his strongest hours the public hardly exists for the writer: he does what he ought to do, has to do, and if afterwards some Public wishes to come and crown him with laurel crowns, well, let it! if critics wish to tell people all that he isn't, well, let them—he knows what he is. But at night when he can't get to sleep it seems to him that it is what he is, his own particular personal quality, that he is being disliked for. It is this that the future will like him for, if it likes him for anything; but will it like him for anything? The poet's hope is in posterity, but it is a pale hope; and now that posterity itself has become a pale hope. . . .

The writer—I am still talking about the writer-not-yet-able-to-go-to-sleep—is willing to have his work disliked, if it's bad; is ready to rest content in dislike, if it's good. But which is it? *He* can't know. He thinks of all those pieces of his that he once thought good, and now thinks bad; how many of his current swans will turn out to be just such ducklings? All of them? If he were worse, would people like him better? If he were better, would people like him worse? If—

He says to himself, "Oh, go to sleep!" And next morning, working at something the new day has brought, he is astonished at the night's thoughts—he does what he does, and lets public, critics, posterity worry about whether it's worth doing. For to tell the truth, the first truth, the poem is a love affair between the poet and his subject, and readers come in only a long time later, as witnesses at the wedding . . .

But what would the ideal witnesses—the ideal public—be? What would an ideal public do? Mainly, essentially, it would just read the poet; read him with a certain willingness and interest; read him imaginatively and perceptively. It needs him, even if it doesn't know that; he needs it, even if he doesn't know that. It and he are like people in one army, one prison, one world: their interests are great and common, and deserve a kind of declaration of dependence. The public might treat him very much as it would like him to treat it. It has its faults, he has his; but both "are, after all," as a man said about women, "the best things that are offered in that line." The public ought not to demand the same old thing from the poet whenever he writes something very new, nor ought it to complain, *The same old thing!* whenever he writes something that isn't very new; and it ought to realize that it is not, unfortunately, in the writer's power to control what he writes: something else originates and controls it, whether you call that something else the unconscious or Minerva or the Muse. The writer writes what he writes just as the public likes what it likes; he can't help himself, it can't help itself, but each of them has to try: most of our morality, most of our culture are in the trying.

We readers can be, or at least can want to be, what the writer himself would want us to be: a public that reads a *lot*—that reads widely, joyfully, and naturally; a public whose taste is formed by acquaintance with the good and great writers of many ages, and not simply acquaintance with a few fashionable contemporaries and the fashionable precursors of those; a public with broad general expectations, but without narrow particular demands, that the new work of art must satisfy; a public that reads with the calm and ease and independence that come from liking things in themselves, for themselves.

This is the kind of public that the poet would like; and if it turned out to be the kind of public that wouldn't like him, why, surely that is something he could bear. It is not his poems but poetry that he wants people to read; if they will read Rilke's and Yeats's and Hardy's poems, he can bear to have his own poems go unread forever. He *knows* that their poems are good to read, and that's something he necessarily can't know about his own; and he knows, too, that poetry itself is good to read —that if you cannot read poetry easily and naturally and joyfully, you are cut off from much of the great literature of the past, some of the good literature of the present. Yet the poet could bear to have people cut off from all that, if only they read widely, naturally, joyfully in the rest of literature: much of the greatest literature, much of the greatest poetry, even, is in prose. If people read this prose—read even a little of it—generously and imaginatively, and felt it as truth and life, as a natural and proper joy, why, that would be enough.

A few months ago I read an interview with a critic; a well-known critic; an unusually humane and intelligent critic. The interviewer had just said that the critic "sounded like a happy man," and the interview was drawing to a close; the critic said, ending it all: "I read, but I don't get time to read at whim. All the reading I do is in order to write or teach, and I resent it. We have no TV, and I don't listen to the radio or records, or go to art galleries or the theater. I'm a completely negative personality."

As I thought of that busy, artless life—no records, no paintings, no plays, no books except those you lecture on or write articles about—I was so depressed that I went back over the interview looking for some bright spot, and I found it, one beautiful sentence: for a moment I had left the gray, dutiful world of the professional critic, and was back in the sunlight and shadow, the unconsidered joys, the unreasoned sorrows, of ordinary readers and writers, amateurishly reading and writing "at whim." The critic said that once a year he read

Kim; and he read *Kim,* it was plain, at whim: not to teach, not to criticize, just for love—he read it, as Kipling wrote it, just because he liked to, wanted to, couldn't help himself. To him it wasn't a means to a lecture or an article, it was an end; he read it not for anything he could get out of it, but for itself. And isn't this what the work of art demands of us? The work of art, Rilke said, says to us always: *You must change your life.* It demands of us that we too see things as ends, not as means—that we too know them and love them for their own sake. This change is beyond us, perhaps, during the active, greedy, and powerful hours of our lives; but during the contemplative and sympathetic hours of our reading, our listening, our looking, it is surely within our power, if we choose to make it so, if we choose to let one part of our nature follow its natural desires. So I say to you, for a closing sentence: *Read at whim! read at whim!*

On Preparing to Read Kipling

MARK TWAIN said that it isn't what they don't know that hurts people, it's what they do know that isn't so. This is true of Kipling. If people don't know about Kipling they can read Kipling, and then they'll know about Kipling: it's ideal. But most people already do know about Kipling—not very much, but too much: they know what isn't so, or what might just as well not be so, it matters so little. They know that, just as Calvin Coolidge's preacher was against sin and the Snake was for it, Kipling was for imperialism; he talked about the white man's burden; he was a crude popular—immensely popular—writer who got popular by writing "If," and "On the Road to Mandalay," and "The Jungle Book," and stories about India like Somerset Maugham, and children's stories; he wrote, "East is East and West is West and never the twain shall meet"; he wrote, "The female of the species is more deadly than the male"—or was that Pope? *Somebody* wrote it. In short: Kipling was someone people used to think was wonderful, but we know better than that now.

People certainly didn't know better than that then. "Dear Harry," William James begins. (It is hard to remember, hard to believe, that anyone ever called Henry James *Harry*, but if it had to be done, William James was the right man to do it.) "Last Sunday I dined with Howells at the Childs', and was delighted to hear him say that you were both a friend and an admirer of Rudyard Kipling. I am ashamed to say that I have been ashamed to write of that infant phenomenon, not knowing, with your exquisitely refined taste, how you might be affected by him and fearing to *jar*. [It is wonderful *to have the engineer/Hoist with his own petard*.] The more rejoiced am I at this, but why didn't you say so ere now? He's more of a Shakespeare than anyone yet in this generation of ours, as it strikes me. And seeing the new effects he lately brings in in "The Light That Failed," and that Simla Ball story with Mrs. Hauksbee in the *Illustrated London News*, makes one sure now that he is only at the beginning of a rapidly enlarging career, with indefinite growth before him. Much of his present coarseness and jerkiness is youth only, divine youth. But *what* a youth! Distinctly the biggest literary phenomenon of our time. He has such human entrails, and he takes less time to get under the heart-strings of his personages than anyone I know. On the whole, bless him.

"All intellectual work is the same,—the artist feeds the public on his own bleeding insides. Kant's *Kritik* is just like a Strauss waltz, and I felt the other day, finishing "The Light That Failed," and an ethical address to be given at Yale College simultaneously, that there was no *essential* difference between Rudyard Kipling and myself as far as that sacrificial element goes."

It surprises us to have James take Kipling so seriously, without reservations, with Shakespeare—to treat him as if he were Kant's *Kritik* and not a Strauss waltz. (Even Henry James, who could refer to "the good little Thomas Hardy"—who was capable of applying to the Trinity itself the adjective *poor*—somehow felt that he needed for Kipling that coarse word *genius*, and called him, at worst, "the great little Kipling.") Similarly, when Goethe and Matthew Arnold write about Byron, we are surprised to see them bringing in Shakespeare—are surprised to see how unquestioningly, with what serious respect, they speak of Byron, as if he were an ocean or a new ice age: "our soul," wrote Arnold, "had *felt* him like the thunder's roll." It is as though mere common sense, common humanity, required this of them: the existence of a world-figure like Byron demands (as the existence of a good or great writer does not) that any inhabitant of the world treat him somehow as the world treats him. Goethe knew that Byron "is a child when he reflects," but this did not prevent him from treating Byron exactly as he treated that other world-figure Napoleon.

An intelligent man said that the world felt Napoleon as a weight, and that when he died it would give a great *oof* of relief. This is just as true of Byron, or of such Byrons of their days as Kipling and Hemingway: after a generation or two the world is tired of being their pedestal, shakes them off with an *oof*, and then—hoisting onto its back a new world-figure—feels the penetrating satisfaction of having made a mistake all its own. Then for a generation or two the Byron lies in the dust where we left him: if the old world did him more than justice, a new one does him less. "If he was so good as all that why isn't he still famous?" the new world asks—if it asks anything. And then when another generation or two are done, we decide that he wasn't altogether a mistake people made in those days, but a real writer after all—that if we like *Childe Harold* a good deal less than anyone thought of liking it then, we like *Don Juan* a good deal more. Byron *was* a writer, people just didn't realize the sort of writer he was. We can feel impatient with Byron's world for liking him for the wrong reasons, and with the succeeding world for disliking him for the wrong reasons, and we are glad that our world, the real world, has at last settled Byron's account.

Kipling's account is still unsettled. Underneath, we still hold it against him that the world quoted him in its sleep, put him in its headlines when he was ill, acted as if he were God; we are glad that we have Hemingway instead, to put in *our* headlines when his plane crashes. Kipling is in the dust, and the dust seems to us a very good place for him. But in twenty or thirty years, when Hemingway is there instead, and we have a new Byron-Kipling-Hemingway to put in our news-programs when his rocket crashes, our resistance to Hemingway will have taken the place of our resistance to Kipling, and we shall find ourselves willing to entertain the possibility that Kipling *was* a writer after all—people just didn't realize the sort of writer he was.

There is a way of travelling into this future—of realizing, now, the sort of writer Kipling was—that is unusually simple, but that people are unusually unwilling to take. The way is: to read Kipling as if one were not prepared to read Kipling; as if one didn't already know about Kipling—had never been told how readers do feel about Kipling, should feel about Kipling; as if one were setting out, naked, to see something that is there naked. I don't entirely blame the reader if he answers: "Thanks very much; if it's just the same to you, I'll keep my clothes on." It's only human of him—man is the animal that wears clothes. Yet aren't works of art in some sense a way of doing without clothes, a means by which reader, writer, and subject are able for once to accept their own nakedness? the nakedness not merely of the "naked truth," but also of the naked wishes that come before and after that truth? To read Kipling, for once, not as the crudely effective, popular writer we know him to be, but as, perhaps, the something else that even crudely effective, popular writers can become, would be to exhibit a magnanimity that might do justice both to

Kipling's potentialities and to our own. Kipling did have, at first, the "coarseness and jerkiness" and mannered vanity of youth, human youth; Kipling did begin as a reporter, did print in newspapers the *Plain Tales from the Hills* which ordinary readers—and, unfortunately, most extraordinary ones—do think typical of his work; but then for half a century he kept writing. Chekhov began by writing jokes for magazines, skits for vaudeville; Shakespeare began by writing *Titus Andronicus* and *The Two Gentlemen of Verona*, some of the crudest plays any crudely effective, popular writer has ever turned out. Kipling is neither a Chekhov nor a Shakespeare, but he is far closer to both than to the clothing-store-dummy-with-the-solar-topee we have agreed to call Kipling. Kipling, like it or not, admit it or not, was a great genius; and a great neurotic; and a great professional, one of the most skillful writers who have ever existed—one of the writers who have used English best, one of the writers who most often have made other writers exclaim, in the queer tone they used for the exclamation: "Well, I've got to admit it really is *written*." When he died and was buried in that foreign land England, that only the Anglo-Indians know, I wish that they had put above his grave, there in *their* Westminster Abbey: "It really was *written*."

Mies van der Rohe said, very beautifully: "I don't want to be interesting, I want to be good." Kipling, a great realist but a greater inventor, could have said that he didn't want to be realistic, he wanted to get it right: that he wanted it not the way it did or —statistics show—does happen, but the way it really would happen. You often feel about something in Shakespeare or Dostoievsky that nobody ever said such a thing, but that it's just the sort of thing people would say if they could—is more real, in some sense, than what people do say. If you have given your imagination free rein, let things go as far as they want to go, the world they made for themselves while you watched can have, for you and later watchers, a spontaneous finality. Some of Kipling has this spontaneous finality; and because he has written so many different kinds of stories—no writer of fiction of comparable genius has depended so much, for so long, on short stories alone —you end dazzled by his variety of realization: so many plants, and so many of them dewy!

If I had to pick one writer to invent a conversation between an animal, a god, and a machine, it would be Kipling. To discover what, if they ever said, the dumb would say—this takes real imagination; and this imagination of what isn't is the extension of a real knowledge of what is, the knowledge of a consummate observer who took no notes, except of names and dates: "if a thing didn't stay in my memory I argued it was hardly worth writing out." Knowing what the peoples, animals, plants, weathers of the world look like, sound like, smell like, was Kipling's *métier*, and so was knowing the words that could make someone else know. You can argue about the judgment he makes of something, but the thing is there. When as a child you first begin to read, what attracts you to a book is illustrations and conversations, and what scares you away is "long descriptions." In Kipling illustration and conversation and description (not long description; read, even the longest of his descriptions is short) have merged into a "toothsome amalgam" which the child reads with a grown-up's ease, and the grown-up with a child's wonder. Often Kipling writes with such grace and command, such a combination of experienced mastery and congenital inspiration, that we repeat with Goethe: "Seeing someone accomplishing arduous things with ease gives us an impression of witnessing the impossible." Sometimes the arduous thing Kipling is accomplishing seems to us a queer, even an absurd thing for anyone to wish to accomplish. But don't we have to learn to consent to this, with Kipling as with other good writers?—to consent to the fact that good writers just don't have good sense; that they are going to write it their way, not ours; that they are never going to have the objective, impersonal rightness they should have, but only the subjective, personal wrongness from which we derived the idea of the rightness. The first thing we notice about *War and Peace* and *Madame Bovary* and *Remembrance of Things Past* is how wonderful they are; the second thing we notice is how much they have wrong with them. They are not at all the perfect work of art we want—so perhaps Ruskin was right when he said that the person who wants perfection knows nothing about art.

Kipling says about a lion cub he and his family had on the Cape: "He dozed on the stoep, I noticed, due north and south, looking with slow eyes up the length of Africa"; this, like several thousand such sentences, makes you take for granted the truth of his "I made my own experiments in the weights, colors, perfumes, and attributes of words in relation to other words, either as read aloud so that they may hold the ear, or, scattered over the page, draw the eye." His words range from gaudy effectiveness to perfection; he is a professional magician but, also, a magician. He says about stories: "A tale from which pieces have been raked out is like a fire that has been poked. One does not know that the operation has been performed, but everyone feels the effect." (He even tells you how best to rake out the pieces: with a brush and Chinese ink you grind yourself.) He is a kind of Liszt—so isn't it just empty bravura, then? Is Liszt's? Sometimes; but sometimes bravura is surprisingly full, sometimes virtuosos are surprisingly plain: to boil a potato perfectly takes a chef home from the restaurant for the day.

Kipling was just such a potato-boiler: a professional knower of professionals, a great trapeze-artist, cabinet-maker, prestidigitator, with all the unnumbered details of others' guilds, crafts, mysteries, techniques at the tip of his fingers—or, at least, at the tip of his tongue. The first sentences he could remember saying as a child had been haltingly translated into English "from the vernacular" (that magical essential phrase for the reader of Kipling!) and just as children feel that it is they and not the grown-ups who see the truth, so Kipling felt about many things that it is the speakers of the vernacular and not the sahibs who tell the truth; that there are many truths that, to be told at all, take the vernacular. From childhood on he learned —to excess or obsession, even—the vernaculars of earth, the worlds inside the world, the many species into which place and language and work divide man. From the species which the division of labor produces it is only a step to the animal species which evolutionary specialization produces, so that Kipling finds it easy to write stories about animals; from the vernaculars or dialects or cants which place or profession produces (Kipling's slogan is, almost, "The cant *is* the man") it is only a step to those which time itself produces, so that Kipling finds it easy to write stories about all the different provinces of the past, or the future (in "As Easy as A.B.C."), or Eternity (if his queer institutional stories of the bureaucracies of Heaven and Hell are located there). Kipling was no Citizen of the World, but like the Wandering Jew he had lived in many places and known many peoples, an uncomfortable stranger repeating to himself the comforts of earth, all its immemorial contradictory ways of being at home.

Goethe, very winningly, wanted to have put on

his grave a sentence saying that he had never been a member of any guild, and was an amateur until the day he died. Kipling could have said, "I never saw the guild I wasn't a member of," and was a professional from the day he first said to his *ayah*, in the vernacular—not being a professional myself, I don't know what it was he said, but it was the sort of thing a man would say who, from the day he was sixteen till the day he was twenty-three, was always —"luxury of which I dream still!"—shaved by his servant before he woke up in the morning.

This fact of his life, I've noticed, always makes hearers give a little shiver; but it is all the mornings when no one shaved Kipling before Kipling woke up, because Kipling had never been to sleep, that make me shiver. "Such night-wakings" were "laid upon me through my life," Kipling writes, and tells you in magical advertising prose how lucky the wind before dawn always was for him. You and I should have such luck! Kipling was a professional, but a professional possessed by both the Daemon he tells you about, who writes some of the stories for him, and the demons he doesn't tell you about, who write some others. Nowadays we've learned to call part of the unconscious *it* or *id;* Kipling had not, but he called this Personal Demon of his *it.* (When he told his father that *Kim* was finished his father asked: "Did *it* stop, or you?" Kipling "told him that it was It.") "When your Daemon is in charge," Kipling writes, "do not try to think consciously. Drift, wait, and obey." He was sure of the books in which "my Daemon was with me . . . When those books were finished they said so themselves with, almost, the water-hammer click of a tap turned off." (Yeats said that a poem finishes itself with a click like a closing box.) Kipling speaks of the "doom of the makers": when their Daemon is missing they are no better than anybody else; but when he is there, and they put down what he dictates, "the work he gives shall continue, whether in earnest or jest." Kipling even "learned to distinguish between the peremptory motions of my Daemon, and the 'carry-over' of induced electricity, which comes of what you might call mere 'frictional' writing." We always tend to distrust geniuses about genius, as if what they say didn't arouse much empathy in us, or as if we were waiting till some more reliable source of information came along; still, isn't what Kipling writes a colored version of part of the plain truth?—there is plenty of supporting evidence. But it is interesting to me to see how thoroughly Kipling manages to avoid any subjective guilt, fallible human responsibility, so that he can say about anything in his stories either: "Entirely conscious and correct, objectively established, independently corroborated, the experts have testified, the professionals agree, it is the consensus of the authorities at the Club," or else: "I had nothing to do with it. I know nothing about it. *It* did it. The Daemon did it all." The reader of Kipling—this reader at least—hates to give all the credit to the Professional or to the Daemon; perhaps the demons had something to do with it too. Let us talk about the demons.

One writer says that we only notice what hurts us—that if you went through the world without hurting anyone, nobody would even know you had been alive. This is quite false, but true, too: if you put it in terms of the derivation of the Principle of Reality from the primary Principle of Pleasure, it does not even sound shocking. But perhaps we only notice a sentence if it sounds shocking—so let me say grotesquely: Kipling was someone who had spent six years in a concentration camp as a child; he never got over it. As a very young man he spent seven years in an India that confirmed his belief in concentration camps; he never got over this either.

As everybody remembers, one of Goya's worst engravings has underneath it: *I saw it*. Some of Kipling has underneath: *It is there*. Since the world is a necessary agreement that it isn't there, the world answered: *It isn't*, and told Kipling what a wonderful imagination he had. Part of the time Kipling answered stubbornly: *I've been there* (*I am there* would have been even truer) and part of the time he showed the world what a wonderful imagination he had. Say *Fairy-tales!* enough to a writer and he will write you fairy-tales. But to our *Are you telling me the truth or are you reassuring yourself?*—we ask it often of any writer, but particularly often of Kipling—he sometimes can say truthfully: *Reassuring you;* we and Kipling have interests in common. Kipling knew that "every nation, like every individual, walks in a vain show—else it could not live with itself"; Kipling knew people's capacity not to see: "through all this shifting, shouting brotheldom the pious British householder and his family bored their way back from the theaters, eyes-front and fixed, as though not seeing." But he himself had seen, and so believed in, the City of Dreadful Night, and the imperturbable or delirious or dying men who ran the city; this City outside was the duplicate of the City inside; and when the people of Victorian Europe didn't believe in any of it, except as you believe in a ghost story, he knew that this was only because they didn't *know*—he knew. So he was obsessed by—wrote about, dreamed about, and stayed awake so as not to dream about—many concentration camps, of the soul as well as of the body; many tortures, hauntings, hallucinations, deliria, diseases, nightmares, practical jokes, revenges, monsters, insanities, neuroses, abysses, forlorn hopes, last chances, extremities of every kind; these and their sweet opposites. He feels the convalescent's gratitude for mere existence, that the world is what the world was: how blue the day is, to the eye that has been blinded! Kipling praises the cessation of pain and its more blessed accession, when the body's anguish blots out for a little "Life's grinning face . . . the trusty Worm that dieth not, the steadfast Fire also." He praises man's old uses, home and all the ways of home: its Father and Mother, there to run to if you could only wake; and praises all our dreams of waking, our fantasies of return or revenge or insensate endurance. He praises the words he has memorized, that man has made from the silence; the senses that cancel each other out, that man has made from the senselessness; the worlds man has made from the world; but he praises and reproduces the sheer charm of—few writers are so purely charming!—the world that does not need to have anything done to it, that is simply there around us as we are there in it. He knows the joy of finding exactly the right words for what there are no words for; the satisfactions of sentimentality and brutality and love too, the "exquisite tenderness" that began in cruelty. But in the end he thanks God most for the small drugs that last—is grateful that He has not laid on us "the yoke of too long Fear and Wonder," but has given us Habit and Work: so that his Seraphs waiting at the Gate praise God

Not for any miracle of easy Loaves and Fishes
But for doing, 'gainst our will, work against our wishes,
Such as finding food to fill daily emptied dishes . . .

praise him

Not for Prophecies or Powers, Visions, Gifts, or Graces
But the unregardful hours that grind us in our places
With the burden on our backs, the weather in our faces.

"Give me the first six years of a child's life and you can have the rest" are the first words of *Some-*

thing of Myself, Kipling's reticent and revealing autobiography. The sentence exactly fits and exactly doesn't fit. For the first six years of his life the child lived in Paradise, the inordinately loved and reasonably spoiled son of the best of parents; after that he lived in the Hell in which the best of parents put him, and paid to have him kept: in "a dark land, and a darker room full of cold, in one wall of which a woman made naked fire . . . a woman who took in children whose parents were in India." The child did not see his parents again for the next six years. He accepted the Hell as "eternally established . . . I had never heard of Hell, so I was introduced to it in all its terrors . . . I was regularly beaten . . . I have known a certain amount of bullying, but this was calculated torture—religious as well as scientific . . . Deprivation from reading was added to my punishments . . . I was well beaten and sent to school through the streets of Southsea with the placard "Liar" between my shoulders . . . Some sort of nervous breakdown followed, for I imagined I saw shadows and things that were not there, and they worried me more than the Woman . . . A man came down to see me as to my eyes and reported that I was half-blind. This, too, was supposed to be 'showing-off,' and I was segregated from my sister—another punishment—as a sort of moral leper."

At the end of the six years the best of parents came back for their leper ("she told me afterwards that when she first came up to my room to kiss me goodnight, I flung up an arm to guard off the cuff I had been trained to expect"), and for the rest of their lives they continued to be the best and most loving of parents, blamed by Kipling for nothing, adored by Kipling for everything: "I think I can truthfully say that those two made up for me the only public for whom then I had any regard whatever till their deaths, in my forty-fifth year."

My *best of parents* cannot help sounding ironic, yet I do not mean it as irony. From the father's bas-reliefs for *Kim* to the mother's "There's no Mother in Poetry, my dear," when the son got angry at her criticism of his poems—from beginning to end they are bewitching; you cannot read about them without wanting to live with them; they were the best of parents. It is *this* that made Kipling what he was: if they had been the worst of parents, even fairly bad parents, even ordinary parents, it would all have made sense, Kipling himself could have made sense out of it. As it was, his world had been torn in two and he himself torn in two: for under the part of him that extenuated everything, blamed for nothing, there was certainly a part that extenuated nothing, blamed for everything—a part whose existence he never admitted, most especially not to himself. He says about some of the things that happened to him during those six years: "In the long run these things and many more of the like drained me of any capacity for real, personal hatred for the rest of my life." To admit from the unconscious something inadmissible, one can simply deny it, bring it up into the light with a *No;* Kipling has done so here—the capacity for real, personal hatred, real, personal revenge, summary fictional justice, is plain throughout Kipling's work. Listen to him tell how he first began to write. He has just been told about Dante: "I bought a fat, American-cloth-bound notebook and set to work on an *Inferno,* into which I put, under appropriate tortures, all my friends and most of the masters." (Why only *most?* Two were spared, one for the Father and one for the Mother.) Succinct and reticent as *Something of Myself* is, it has room for half a dozen scenes in which the helpless Kipling is remorselessly, systematically, comprehensively humiliated before the inhabitants of his universe. At school, for instance: "H—— then told me off before my delighted companions in his best style, which was acid and contumelious. He wound up with a few general remarks about dying as a 'scurrilous journalist' . . . The tone, matter, and setting of his discourse were as brutal as they were meant to be—brutal as the necessary wrench on the curb that fetches up a too-flippant colt." Oh, necessary, entirely necessary, we do but torture in education! one murmurs to these methodical justifications of brutality as methodical, one of authority's necessary stages. Here is another master: "Under him I came to feel that words could be used as weapons, for he did me the honor to talk at me plentifully . . . One learns more from a good scholar in a rage than from a score of lucid and laborious drudges; and to be made the butt of one's companions in full form is no bad preparation for later experiences. I think this 'approach' is now discouraged for fear of hurting the soul of youth, but in essence it is no more than rattling tins or firing squibs under a colt's nose. I remember nothing save satisfaction or envy when C—— broke his precious ointments over my head." Nothing? Better for Kipling if he had remembered—not remembering gets rid of nothing. Yet who knows? he may even have felt—known that he felt—"nothing save satisfaction and envy," the envying satisfaction of identification. As he says, he was learning from a master to use words as weapons, but he had already learned from his life a more difficult lesson: to know that, no matter how the sick heart and raw being rebel, it is all for the best; in the past there were the best of masters and in the future there will be the best of masters, if only we can wait out, bear out, the brutal present—the incomprehensible present that some day we shall comprehend as a lesson.

The scene changes from England to India, school to Club, but the action—passion, rather—is the same: "As I entered the long, shabby dining-room where we all sat at one table, everybody hissed. I was innocent enough to ask: 'What's the joke? Who are they hissing?' 'You,' said the man at my side. 'Your damn rag has ratted over the Bill.' It is not pleasant to sit still when one is twenty while all your universe hisses you." One expects next a sentence about how customary and salutary hissing is for colts, but for once it doesn't come; and when

Kipling's syntax suffers as it does in this sentence, he is remembering something that truly is not pleasant. He even manages somewhat to justify, somehow to justify, his six years in Hell: the devils' inquisitions, after all, "made me give attention to the lies I soon found it necessary to tell; and this, I presume, is the foundation of literary effort. . . Nor was my life an unsuitable preparation for my future, in that it demanded constant wariness, the habit of observation and attendance on moods and tempers; the noting of discrepancies between speech and action; a certain reserve of demeanor; and automatic suspicion of sudden favors." I have seen writers called God's spies, but Kipling makes it sound as if they were just spies—or spies on God. If only he could have blamed God—his Gods—a little consciously, forgiven them a little unconsciously! could have felt that someone, sometimes, doesn't *mean* something to happen! But inside, and inside stories, everything is meant.

After you have read Kipling's fifty or seventy-five best stories you realize that few men have written this many stories of this much merit, and that very few have written more and better stories. Chekhov and Turgenev are two who immediately come to mind; and when I think of their stories I cannot help thinking of what seems to me the greatest lack in Kipling's. I don't know exactly what to call it: a lack of dispassionate moral understanding, perhaps—of the ability both to understand things and to understand that there is nothing to do about them. (In a story, after all, there is always something you *can* do, something that a part of you is always trying to make you do.) Kipling is a passionate moralist, with a detailed and occasionally profound knowledge of part of things; but his moral spectrum has shifted, so that he can see far down into the infra-red, but is blind for some frequencies normal eyes are sensitive to. His morality is the one-sided, desperately protective, sometimes vindictive morality of someone who has been for some time the occupant of one of God's concentration camps, and has had to spend the rest of his life justifying or explaining out of existence what he cannot forget. Kipling tries so hard to celebrate and justify true authority, the work and habit and wisdom of the world, because he feels so bitterly the abyss of pain and insanity that they overlie, and can do—even will do—nothing to prevent.

Kipling's morality is the morality of someone who has to prove that God is not responsible for part of the world, and that the Devil is. If Father and Mother were not to blame for anything, yet what did happen to you could happen to you—if God is good, and yet the concentration camps exist —then there has to be *someone* to blame, and to punish too, some real, personal source of the world's evil. (He finishes "At the End of the Passage" by having someone quote: "There may be Heaven, there must be Hell./ Meanwhile there is our life here. Well?" In most of his stories he sees to it that our life here is Heaven and Hell.) But in this world, often, there is nothing to praise but no one to blame, and Kipling can bear to admit this in only a few of his stories. He writes about one source of things in his childhood: "And somehow or other I came across a tale about a lion-hunter in South Africa who fell among lions who were all Freemasons, and with them entered into a conspiracy against some wicked baboons. I think that, too, lay dormant until the Jungle Books began to be born." In Chekhov or Turgenev, somehow or other, the lions aren't really Freemasons and the baboons aren't really wicked. In Chekhov and Turgenev, in fact, most of the story has disappeared from the story: there was a lion-hunter in South Africa, and first he shot the lions, and then he shot the baboons, and finally he shot himself; and yet it wasn't *wicked*, exactly, but human—very human.

Kipling had learned too well and too soon that, in William James' words: "The normal process of life contains moments as bad as any of those which insane melancholy is filled with, moments in which radical evil gets its innings and takes its solid turn. The lunatic's visions of horror are all drawn from the material of daily fact. Our civilization is founded on the shambles, and each individual existence goes out in a lonely spasm of helpless agony. If you protest, my friend, wait till you arrive there yourself!" Kipling had arrived there early and returned there often. One thinks sadly of how deeply congenial to this torturing obsessive knowledge of

Kipling's the first World War was: the death and anguish of Europe produced some of his best and most terrible stories, and the death of his own son, his own anguish, produced "Mary Postgate," that nightmare-ish, most human and most real day-dream of personal revenge. The world *was* Hell and India underneath, after all; and he could say to the Victorian, Edwardian Europeans who had thought it all just part of his style: "You wouldn't believe me!"

Svidrigaylov says: "We are always thinking of eternity as an idea that cannot be understood, something immense. But why must it be? What if, instead of all this, you suddenly find just a little room there, something like a village bath-house, grimy, and spiders in every corner, and that's all eternity is . . . I, you know, would certainly have made it so deliberately." Part of Kipling would have replied to this with something denunciatory and biblical, but another part would have blurted eagerly, like somebody out of *Kim:* "Oah yess, that is dam-well likely! Like a dak-bungalow, you know." It is an idea that would have occurred to him, down to the last *deliberately*.

But still another part of Kipling would suddenly have seen—he might even later have written it down, according to the dictates of his Daemon—a story about a boy who is abandoned in a little room, grimy, with spiders in every corner, and after a while the spiders come a little nearer, and a little nearer, and one of them is Father Spider, and one of them is Mother Spider, and the boy is their Baby Spider. To Kipling the world was a dark forest full of families: so that when your father and mother leave you in the forest to die, the wolves that come to eat you are always Father Wolf and Mother Wolf, your real father and real mother, and you are—as not even the little wolves ever quite are—their real son. The family romance, the two families of the Hero, have so predominant a place in no other writer. Kipling never said a word or thought a thought against his parents, "both so entirely comprehending that except in trivial matters we had hardly need of words"; few writers have made authority so tender, beautiful, and final—have had us miserable mortals serve better masters; *but* Kipling's Daemon kept bringing Kipling stories in which wild animals turn out to be the abandoned Mowgli's real father and mother, a heathen Lama turns out to be the orphaned Kim's real father—and Kipling wrote down the stories and read them aloud to his father and mother.

This is all very absurd, all very pathetic? Oh yes, that's very likely; but, reader, down in the darkness where the wishes sleep, snuggled together like bats, you and I are Baby Spider too. If you think *this* absurd you should read Tolstoy—all of Tolstoy. But I should remark, now, on something that any reader of Kipling will notice: that though he can seem extraordinarily penetrating or intelligent—inspired, even—he can also seem very foolish or very blind. This is a characteristic of the immortals from which only we mortals are free. They over-say everything. It is only ordinary readers and writers who have ordinary common sense, who are able to feel about things what an ordinarily sensible man should. To another age, of course, our ordinary common sense will seem very very common and ordinary, but not sense, exactly: sense never lasts for long; instead of having created our own personal day-dream or nightmare, as the immortals do, we merely have consented to the general day-dream or nightmare which our age accepted as reality—it will seem to posterity only sense to say so, and it will say so, before settling back into a common sense of its own.

In the relations of mortals and immortals, yesterday's and today's posterities, there is a certain pathos or absurdity. There is a certain absurdity in my trying to persuade you to read Kipling sympathetically—who are *we* to read or not read Kipling sympathetically? part of me grunts. Writing about just which writers people are or are not attracted to, these years— who was high in the 19th, who's low in the 20th—all the other stock-market quotations of the centuries, makes me feel how much such things have to do with history, and how little with literature. The stories themselves are literature. While their taste is on my tongue, I can't help feeling that virtue is its own reward, that good writing will take care of itself. It is a feeling I have often had after reading all of an author: that there it is. I can see that if I don't write this about the stories, plenty of other writers will; that if you don't read the stories, plenty of other readers will. The man Kipling, the myth Kipling is over; but the stories themselves—Kipling—have all the time in the world. The stories—some of them—can say to us with the calm of anything that has completely realized its own nature: "Worry about yourselves, not us. *We're* all right."

And yet, I'd be sorry to have missed them, I'd be sorry for you to miss them. I have read one more time what I've read so often before, and have picked for you what seem—to a loving and inveterate reader, one ashamed of their faults and exalted by their virtues—fifty of Kipling's best stories.

Stories

STORY, the dictionary tells one, is a short form of the word *history*, and stands for *a narrative, recital, or description of what has occurred;* just as it stands for *a fictitious narrative, imaginative tale; Colloq. a lie, a falsehood.*

A story, then, tells the truth or a lie—is a wish, or a truth, or a wish modified by a truth. Children ask first of all: "Is it a *true* story?" They ask this of the storyteller, but they ask of the story what they ask of a dream: that it satisfy their wishes. The Muses are the daughters of hope and the stepdaughters of memory. The wish is the first truth about us, since it represents not that learned principle of reality which half-governs our workaday hours, but the primary principle of pleasure which governs infancy, sleep, daydreams—and, certainly, many stories. Reading stories, we cannot help remembering Groddeck's "We have to reckon with what exists, and dreams, daydreams too, are also facts; if anyone really wants to investigate realities, he cannot do better than to start with such as these.

If he neglects them, he will learn little or nothing of the world of life." If wishes were stories, beggars would read; if stories were true, our saviors would speak to us in parables. Much of our knowledge of, our compensation for, "the world of life" comes from stories; and the stories themselves are part of "the world of life." Shakespeare wrote:

This is an art
Which does mend nature, change it rather, but
The art itself is nature . . .

and Goethe, agreeing, said: "A work of art is just as much a work of nature as a mountain."

In showing that dreams sometimes both satisfy our wishes and punish us for them, Freud compares the dreamer to the husband and wife in the fairy tale of The Three Wishes: the wife wishes for a pudding, the husband wishes it on the end of her nose, and the wife wishes it away again. A contradictory family! But it is this family—wife, husband, and pudding—which the story must satisfy: the writer is, and is writing for, a doubly- or triply-natured creature, whose needs, understandings, and ideals—whether they are called id, ego, and superego, or body, mind, and soul—contradict one another. Most of the stories that we are willing to call works of art are compounds almost as complicated as their creators; but occasionally we can see isolated, in naked innocence, one of the elements of which our stories are composed. Thomas Leaf's story (in Hardy's *Under the Greenwood Tree*) is an example:

"Once," said the delighted Leaf, in an uncertain voice, "there was a man who lived in a house! Well, this man went thinking and thinking night and day. At last, he said to himself, as I might, 'If I had only ten pound, I'd make a fortune.' At last by hook or by crook, behold he got the ten pounds!"

"Only think of that!" said Nat Callcome satirically.

"Silence!" said the tranter.

"Well, now comes the interesting part of the story! in a little time he made that ten pounds twenty. Then a little after that he doubled it, and made it forty. Well, he went on, and a good while after that he made it eighty, and on to a hundred. Well, by-and-by he made it two hundred! Well, you'd never believe it, but —he went on and made it four hundred! He went on, and what did he do? Why, he made it eight hundred! Yes, he did," continued Leaf, in the highest pitch of excitement, bringing down his fist upon his knee, with such force that he quivered with the pain; "yes, and he went on and made it A THOUSAND!"

"Hear, hear!" said the tranter. "Better than the history of England, my sonnies!"

"Thank you for your story, Thomas Leaf," said grandfather William; and then Leaf gradually sank into nothingness again.

Every day, in books, magazines, and newspapers, over radio and television, in motion-picture theaters, we listen to Leaf's story one more time, and then sink into nothingness again. His story is, in one sense, better than the history of England—or would be if the history of England were not composed, among other things, of Leaf's story and a million like it. His story, stood on its head, is the old woman's story in *Wozzeck*. "Grandmother, tell us a story," beg the children. "All right, you little crabs," she answers.

Once upon a time there was a poor little girl who had no father and mother because everyone was dead and there was no one left in the whole world. Everyone was dead, and she kept looking for someone night and day. And since there was no one on earth, she thought she'd go to heaven. The moon looked at her so friendly, but when she finally got to it, it was just a piece of rotted wood. So she went on to the sun, and when she got there, it was just a dried-up sunflower. And when she got to the stars, they were just little gold flies stuck up there as if they'd been caught in a spider web. And when she thought she'd go back to earth, it was just an upside-down pot. And she was all alone. And so she sat down and cried. And she's still sitting there, all alone.

The grandmother's story is told less often—but often enough: when we wake into the reality our dream has contradicted, we are bitter at returning against our wishes to so bad a world, and take a fierce pleasure in what remains to us, the demonstration that it is the worst of all possible worlds. And we take pleasure also—as our stories show—in repeating over and over, until we can bear it, all that we found unbearable: the child whose mother left her so often that she invented a game of throwing her doll out of her crib, exclaiming as it vanished: "Gone! gone!" was a true poet. "Does I 'member much about slavery times?" the old man says, in *Lay My Burden Down;* "well, there is no way for me to disremember unless I die." But the worst memories are joyful ones: "Every time Old Mistress thought we little black children was hungry 'tween meals she would give us johnnycake and plenty of buttermilk to drink with it. There was a long trough for us they would scrub so clean. They would fill this trough with buttermilk and all us children would sit round the trough and drink with our mouths and hold our johnnycake with our hands. I can just see myself drinking now. It was so good . . ." It is so good, our stories believe, simply to remember: their elementary delight in recognition, familiarity, mimesis, is another aspect of their obsession with all the likenesses of the universe, those metaphors that Proust called essential to style. Stories want to *know:* everything from the first blaze and breathlessness and fragrance to the last law and structure; but, too, stories don't want to know, don't want to care, just want to *do as they please.* (Some great books are a consequence of the writer's losing himself in his subject, others are a consequence of his losing himself in himself. Rabelais' "do what you please" is the motto of how many masterpieces, from Cervantes and Sterne on up to the present.) For stories vary from a more-than-Kantian disinterestedness, in which the self is a representative, indistinguishable integer among millions—the mere *one* or *you* or *man* that is the subject of all the verbs—to an insensate, protoplasmic egotism to which the self is the final fact, a galaxy that it is impracticable to get out of to other galaxies. Polarities like these are almost the first thing one notices about fiction. It is as much haunted by the chaos which precedes and succeeds order as by order; by the incongruities of the universe (wit, humor, the arbitrary, accidental, and absurd—all irruptions from, releases of, the unconscious) as by its likenesses. A story may present fantasy as fact, as the sin or *hubris* that the fact of things punishes, or as a reality superior to fact. And, often, it presents it as a mixture of the three: all opposites meet in fiction.

The truths that he systematized, Freud said, had already been discovered by the poets; the tears of things, the truth of things, are there in their fictions. And yet, as he knew, the root of all stories is in Grimm, not in La Rochefoucauld; in dreams, not in cameras and tape recorders. Turgenev was right when he said, "Truth alone, however powerful, is not art"—oxygen alone, however concentrated, is not water; and Freud was right, profoundly right, when he showed "that the dream is a compromise between the expression of and the defence against the unconscious emotions; that in it the unconscious wish is represented as being fulfilled; that there are very definite mechanisms that control this expression; that the primary process controls the dream world just as it controls the entire unconscious life of the soul, and that myth and poetical productions come into being in the same way and have the same meaning. There is only one important difference: in the myths and in the works of poets the secondary elaboration is much further developed, so that a logical and coherent entity is created." It is hard to exaggerate the importance of this difference, of course; yet usually we do exaggerate it—do write as if that one great difference had hidden from us the greater similarities which underlie it.

II

A BABY ASLEEP but about to be waked by hunger sometimes makes little sucking motions: he is dreaming that he is being fed, and manages by virtue of the dream to stay asleep. He may even smile a little in satisfaction. But the smile cannot last for long—the dream fails, and he wakes. This is, in a sense, the first story; the child in his "impotent omnipotence" is like us readers, us writers, in ours.

A story is a chain of events. Since the stories that we know are told by men, the events of the story happen to human or anthropomorphic beings—gods, beasts, and devils, and are related in such a way that the story seems to begin at one place and to end at a very different place, without any essential interruption to its progress. The poet or storyteller, so to speak, writes numbers on a blackboard, draws a line under them, and adds them into their true but unsuspected sum. Stories, because of their nature or—it is to say the same thing—of ours, are always capable of generalization: a story about a dog Kashtanka is true for all values of dogs and men.

Stories can be as short as a sentence. Bion's saying, *The boys throw stones at the frogs in sport, but the frogs die not in sport but in earnest,* is a story; and when one finds it in Aesop, blown up into a fable five or six sentences long, it has become a poorer story. Blake's *Prudence is a rich, ugly old maid courted by Incapacity* has a story inside it, waiting to flower in a glass of water. And there is a story four sentences long that not even Rilke was able to improve: *Now King David was old and stricken in years; and they covered him with clothes, but he got no heat. Wherefore his servants said unto him, Let there be sought for my lord the king a young virgin: and let her stand before the king, and let her cherish him, and let her lie in thy bosom, that my lord the king may get heat. So they sought for a fair damsel throughout all the coasts of Israel, and found Abishag a Shunamite, and brought her to the king. And the damsel was very fair, and cherished the king, and ministered to him: but the king knew her not. . . .* The enlisted men at Fort Benning buried their dog Calculus under a marker that read: *He made better dogs of us all;* and a few days ago I read in the paper: *A Sunday-school teacher, mother of four children, shot to death her eight-year-old daughter as she slept today, state police reported. Hilda Kristle, 43, of Stony Run, told police that her youngest daughter, Suzanne, "had a heavy heart and often went about the house sighing."*

When we try to make, out of these stories life gives us, works of art of comparable concision, we almost always put them into verse. Blake writes about a chimney sweep:

A little black thing among the snow
Crying "'weep! 'weep!" in notes of woe!
"Where are thy father & mother, say?"
"They are both gone up to the church to pray.

"Because I was happy upon the heath,
And smil'd among the winter's snow
They clothed me in the clothes of death,
And taught me to sing the notes of woe.

"And because I am happy & dance & sing,
They think they have done me no injury,
And are gone to praise God & his Priest & King,
Who make up a Heaven of our misery—"

and he has written enough. Stephen Crane says in fifty words:

In the desert
I saw a creature naked, bestial,
Who, squatting upon the ground,
Held his heart in his hands
And ate of it.
I said, "Is it good, friend?"
"It is bitter—bitter," he answered;
"But I like it
Because it is bitter
And because it is my heart."

These are the bones of stories, and we shiver at them. The poems one selects for a book of stories have more of the flesh of ordinary fiction. A truly representative book of stories would include a great many poems: during much of the past people put into verse the stories that they intended to be literature.

But it is hard to put together any representative collection of stories. It is like starting a zoo in a closet: the giraffe alone takes up more space than one has for the collection. *Remembrance of Things Past* is a story, just as Saint-Simon's memoirs are a great many stories. One can represent the memoirs with the death of Monseigneur, but not even the death of Bergotte, the death of the narrator's grandmother, can do that for *Remembrance of Things Past.* Almost everything in the world, one realizes after a while, is too long to go into a short book of stories—a book of short stories. So, even, are all those indeterminate masterpieces that the nineteenth century called short stories and that we call short novels or novelettes: Tolstoy's *The Death of Ivan Ilyich*, *Hadji Murad*, *Master and Man;* Flaubert's *A Simple Heart;* Mann's *Death in Venice;* Leskov's *The Lady Macbeth of the Mzinsk District;* Keller's *The Three Righteous Comb-Makers;* James's *The Aspern Papers;* Colette's *Julie de Carneilhan;* Kleist's *Michael Kohlhaas;* Joyce's *The Dead;* Turgenev's *A Lear of the Steppes;* Hofmannsthal's *Andreas;* Kafka's *Metamorphosis;* Faulkner's *Spotted Horses;* Porter's *Old Mortality;* Dostoievsky's *The Eternal Husband;* Melville's *Bartleby the Scrivener*, *Benito Cereno;* Chekhov's *Ward No. 6*, *A Dreary Story*, *Peasants*, *In the Ravine*.

And there are many more sorts of stories than

there are sizes. Epics; ballads; historical or biographical or autobiographical narratives, letters, diaries; myths, fairy tales, fables; dreams, daydreams; humorous or indecent or religious anecdotes; all those stories which might be called specialized or special case—science fiction, ghost stories, detective stories, Westerns, True Confessions, children's stories, and the rest; and, finally, "serious fiction"—Proust and Chekhov and Kafka, *Moby-Dick, Great Expectations, A Sportsman's Notebook*. What I myself selected for a book of stories was most of it "serious fiction," some of it serious fiction in verse; but there was a letter of Tolstoy's, a piece of history and autobiography from Saint-Simon; and there were gipsy and German fairy tales, Hebrew and Chinese parables, and two episodes from the journal of an imaginary Danish poet, the other self of the poet Rainer Maria Rilke. There are so many good short narratives of every kind that a book of three or four hundred pages leaves almost all of their writers unrepresented. By saying that I was saving these writers for a second and third book I tried to make myself feel better at having left them out of the first. For I left out all sagas, all ballads, all myths; a dozen great narrators in verse, from Homer to Rilke; Herodotus, Plutarch, Pushkin, Hawthorne, Flaubert, Dostoievsky, Melville, James, Leskov, Keller, Kipling, Mann, Faulkner—I cannot bear to go on. Several of these had written long narratives so much better than any of their short ones that it seemed unfair to use the short, and it was impossible to use the long. Hemingway I could not get permission to reprint. Any anthology is, as the dictionary says, a bouquet—a bouquet that leaves out most of the world's flowers.

My own bunch is named *The Anchor Book of Stories*, and consists of Franz Kafka's *A Country Doctor;* Anton Chekhov's *Gusev;* Rainer Maria Rilke's *The Wrecked Houses* and *The Big Thing* (from *The Notebooks of Malte Laurids Brigge*); Robert Frost's *The Witch of Coös;* Giovanni Verga's *La Lupa;* Nicolai Gogol's *The Nose;* Elizabeth Bowen's *Her Table Spread;* Ludwig Tieck's *Fair Eckbert;* Bertolt Brecht's *Concerning the Infanticide Marie Farrar;* Lev Tolstoy's *The Three Hermits;* Peter Taylor's *What You Hear from 'Em?;* Hans Christian Andersen's *The Fir Tree;* Katharine Anne Porter's *He;* a Gipsy's *The Red King and the Witch;* Anton Chekhov's *Rothschild's Fiddle;* the Brothers Grimm's *Cat and Mouse in Partnership;* E. M. Forster's *The Story of the Siren; The Book of Jonah;* Franz Kafka's *The Bucket-Rider;* Saint-Simon's *The Death of Monseigneur;* Isaac Babel's *Awakening;* five anecdotes by Chuang T'zu; Hugo von Hofmannsthal's *A Tale of the Cavalry;* William Blake's *The Mental Traveller;* D. H. Lawrence's *Samson and Delilah;* Lev Tolstoy's *The Porcelain Doll;* Ivan Turgenev's *Byezhin Prairie;* William Wordsworth's *The Ruined Cottage;* Frank O'Connor's *Peasants;* and Isak Dinesen's *Sorrow-Acre.*

I disliked leaving out writers, but I disliked almost as much having to leave out some additional stories by some of the writers I included. I used so many of the writers who "came out of Gogol's *Overcoat*" that *The Overcoat* was in a sense already there, but I wished that it and *Old-World Landowners* had been there in every sense; that I could have included Chekhov's *The Bishop, The Lady with the Dog, Gooseberries, The Darling, The Man in a Shell, The Kiss, The Witch, On Official Business*, and how many more; that I could have included Kafka's *The Penal Colony* and *The Hunter Gracchus;* and that I could have included at least a story more from Lawrence, Tolstoy, Verga, Grimm, and Andersen. With Turgenev's masterpiece all selection fails: *A Sportsman's Notebook* is a whole greater and more endearing than even the best of its parts.

III

THERE ARE all kinds of beings, and all kinds of things happen to them; and when you add to these what are as essential to the writer, the things that don't actually happen, the beings that don't actually exist, it is no wonder that stories are as varied as they are. But it seems to me that there are two extremes: stories in which nothing happens, and stories in which everything is a happening. The Muse of fiction believes that people "don't go to the North Pole" but go to work, go home, get married, die; but she believes at the same time that absolutely anything can occur—concludes, with Gogol: "Say what you like, but such things do happen —not often, but they do happen." Our lives, even our stories, approach at one extreme the lives of Prior's Jack and Joan:

If human things went Ill or Well;
If changing Empires rose or fell;
The Morning past, the Evening came,
And found this couple still the same.
They Walked and Eat, good folks: What then?
Why then they Walk'd and Eat again:
They soundly slept the Night away:
They did just Nothing all the day . . .
Nor Good, nor Bad, nor Fools, nor Wise;
They wou'd not learn, nor cou'd advise:
Without Love, Hatred, Joy, or Fear,
They led—a kind of—as it were;
Nor Wish'd, nor Car'd, nor Laugh'd, nor Cry'd:
And so They liv'd; and so They dy'd.

Billions have lived, and left not even a name behind, and while they were alive nobody knew their names either. These live out their lives "among the rocks and winding scars/ Where deep and low the hamlets lie/ Each with its little patch of sky/ And little lot of stars"; soundly sleep the Night away in the old houses of Oblomov's native village, where everybody did just Nothing all the day; rise—in Gogol's Akaky Akakyevich Bashmachkin, in the *Old-World Landowners,* to a quite biblical pathos and grandeur; are relatives of that Darling, that *dushechka,* who for so many solitary years "had no opinions of any sort. She saw the objects about her and understood what she saw, but could not form any opinion about them"; sit and, "musing with close lips and lifted eyes/ Have smiled with self-contempt to live so wise/ And run so smoothly such a length of lies"; walk slowly, staring about them—or else just walk—through the pages of Turgenev, Sterne, Keller, Rabelais, Twain, Cervantes, and how many others; and in Chuang T'zu disappear into the mists of time, looming before us in primordial grandeur: "In the days of Ho Hsu the people did nothing in particular when at rest, and went nowhere in particular when they moved. Having food, they rejoiced; having full bellies, they strolled about. Such were the capacities of the people."

How different from the later times, the other pages, in which people "wear the hairs off their legs" "counting the grains of rice for a rice-pudding"! How different from the other extreme: the world of Svidrigaylov, Raskolnikov, Stavrogin, where everything that occurs is either a dream told as if it were reality, or reality told as if it were a dream, and where the story is charged up to the point at which the lightning blazes out in some nightmare, revelation, atrocity, and the drained narrative can begin to charge itself again. In this world, and in the world of *The Devil*, *The Kreutzer Sonata*, *The Death of Ivan Ilyich*, everything is the preparation for, or consummation of, an Event; everyone is an echo of "the prehistoric, unforgettable Other One, who is never equalled by anyone later." This is the world of Hofmannsthal's *A Tale of the Cavalry*, where even the cow being dragged to the shambles, "shrinking from the smell of blood and the fresh hide of a calf nailed to the doorpost, planted its hooves firm on the ground, drew the reddish haze of the sunset in through dilated nostrils, and, before the lad could drag her across the road with stick and rope, tore away with piteous eyes a mouthful of the hay which the sergeant had tied on the front of his saddle." It is the world of Nijinsky's diary: "One evening I went for a walk up the hill, and stopped on the mountain . . . 'the mountain of Sinai.' I was cold. I had walked far. Feeling that I should kneel, I quickly knelt and then felt that I should put my hand in the snow. After doing this, I suddenly felt a pain and cried with it, pulling my hand away. I looked at a star, which did not say good evening to me. It did not twinkle at me. I got frightened and wanted to run, but could not because my knees were rooted to the snow. I started to cry, but no one heard my weeping. No one came to my rescue. After several minutes I turned and saw a house. It was closed and the windows shuttered . . . I felt frightened and shouted at the top of my voice: 'Death!' I did not know why, but felt that one must shout 'Death!' After that I felt warmer . . . I walked on the snow which crunched beneath my feet. I liked the snow and listened to its crunching. I loved listening to my footsteps; they were full of life. Looking at the sky, I saw the stars which were twinkling at me and felt merriment in them. I was happy and no longer felt cold . . . I started to go down a dark road, walking quickly, but was stopped by a tree which saved me. I was on the edge of a precipice. I thanked the tree. It felt me because I caught hold of it; it received my warmth and I received the warmth of the tree. I do not know who most needed the warmth. I walked on and suddenly stopped, seeing a precipice without a tree. I understood that God had stopped me because He loves me, and therefore said: 'If it is thy will, I will fall down the precipice. If it is thy will, I will be saved.' "

This is what I would call pure narrative; one must go to writers like Tolstoy and Rilke and Kafka to equal it. In the unfinished stories of Kafka's notebook, some fragment a page long can carry us over a whole abyss of events: "I was sitting in the box, and beside me was my wife. The play being performed was an exciting one, it was about jealousy; at that moment in the midst of a brilliantly lit hall surrounded by pillars, a man was just raising his dagger against his wife, who was slowly retreating to the exit. Tense, we leaned forward together over the balustrade; I felt my wife's curls against my temples. Then we started back, for something moved on the balustrade; what we had taken for the plush upholstery of the balustrade was the back of a tall thin man, not an inch broader than the balustrade, who had been lying flat on his face there and was now slowly turning over as though trying to find a more comfortable position. Trembling, my wife clung to me. His face was quite close to me, narrower than my hand, meticulously clean as that of a waxwork figure, and with a pointed black beard. 'Why do you come and frighten us?' I exclaimed. 'What are you up to here?' 'Excuse me!' the man said, 'I am an admirer of your wife's. To feel her elbows on my body makes me happy.' 'Emil, I implore you, protect me!' my wife exclaimed. 'I too am called Emil,' the man said, supporting his head on one hand and lying there as though on a sofa. 'Come to me, dear sweet little woman.' 'You cad,' I said, 'another word and you'll find yourself lying down there in the pit,' and as though certain that this word was bound to come, I tried to push him over, but it was not so easy, he seemed to be a solid part of the balustrade, it was as though he were built into it, I tried to roll him off, but I couldn't do it, he only laughed and said: 'Stop that, you silly little man, don't wear out your strength prematurely, the struggle is only beginning and it will end, as may well be said, with your wife's granting my desire.' 'Never!' my wife exclaimed, and then, turning to me: 'Oh, please, do push him down now.' 'I can't,' I exclaimed, 'you can see for yourself how I'm straining, but there's some trickery in it, it can't be done.' 'Oh dear, oh dear,' my wife lamented, 'what is to become of me?' 'Keep calm,' I said, 'I beg of you. By getting so worked up you're only making it worse, I have another plan now, I shall cut the plush open here with my knife and then drop the whole thing down and the fellow with it.' But now I could not find my knife. 'Don't you know where I have my knife?' I asked. 'Can I have left it in my overcoat?' I was almost going to dash along to the cloakroom when my wife brought me to my senses. 'Surely you're not going to leave me alone now, Emil,' she cried. 'But if I have no knife,' I shouted back. 'Take mine,' she said and began fumbling in her little bag, with trembling fingers, but then of course all she produced was a tiny little mother-of-pearl knife."

One of the things that make Kafka so marvellous a writer is his discovery of—or, rather, discovery by—a kind of narrative in which logical analysis and humor, the greatest enemies of narrative movement, have themselves become part of the movement. In narrative at its purest or most eventful we do not understand but are the narrative. When we understand completely (or laugh completely, or feel completely a lyric empathy with the beings of the world), the carrying force of the narrative is dissipated: in fiction, to understand everything is to get nowhere. Yet, walking through Combray with Proust, lying under the leaves with Turgenev and the dwarf Kasyan, who has ever wanted to get anywhere but where he already is, in the best of all possible places?

In stories-in-which-everything-is-a-happening each event is charged and about to be further charged, so that the narrative may at any moment reach a point of unbearable significance, and disintegrate into energy. In stories-in-which-nothing-happens even the climax or denouement is liable to lose what charge it has, and to become simply one more portion of the lyric, humorous, or contemplative continuum of the story: in Gogol's *The Nose* the policeman seizes the barber, the barber turns pale, "but here the incident is completely shrouded in a fog and absolutely nothing is known of what happened next"; and in *Nevsky Avenue*, after Schiller, Hoffman, and Kuntz the carpenter have stripped Lieutenant Pirogov and "treated him with such an utter lack of ceremony that I cannot find words to describe this highly regrettable incident," Pirogov goes raging away, and "nothing could compare with Pirogov's anger and indignation. Siberia and the lash seemed to him the least punishment Schiller deserved . . . But the whole thing somehow petered out most strangely: on the way to the general, he went into a pastry-cook's, ate two pastries, read something out of the *Northern Bee*, and left with his anger somewhat abated"; took a stroll along Nevsky Avenue; and ended at a party given by one of the directors of the Auditing Board, where he "so distinguished himself in the mazurka that not only the ladies but also the gentlemen were in raptures over it. What a wonderful world we live in!"

One of these extremes of narrative will remind us of the state of minimum excitation which the organism tries to re-establish—of the baby asleep, a lyric smile on his lips; the other extreme resembles the processes of continually increased excitation found in sex and play.

The Woman at the Washington Zoo

CRITICS *fairly often write essays about how some poem was written; the poet who wrote it seldom does. When Robert Penn Warren and Cleanth Brooks were making a new edition of* Understanding Poetry, *they asked several poets to write such essays. I no longer remembered much about writing* The Woman at the Washington Zoo *—a poem is, so to speak, a way of making you forget how you wrote it—but I had almost all the sheets of paper on which it was written, starting with a paper napkin from the Methodist Cafeteria. If you had asked me where I had begun the poem I'd have said: "Why, Sir, at the beginning"; it was a surprise to me to see that I hadn't.*

As I read, arranged, and remembered the pages it all came back to me. I went over them for several days, copying down most of the lines and phrases and mentioning some of the sights and circumstances they came out of; I tried to give a fairly good idea of the objective process of writing the poem. You may say, "But isn't a poem a kind of subjective process, like a dream? Doesn't it come out of unconscious wishes of yours, childhood memories, parts of your own private emotional life?" It certainly does: part of them I don't know about and the rest I didn't write about. Nor did I write about or copy down something that begins to appear on the last two or three pages: lines and phrases from a kind of counter-poem, named Jerome, *in which St. Jerome is a psychoanalyst and his lion is at the zoo.*

If after reading this essay the reader should say: "You did all that you could to the things, but the things just came," he would feel about it as I do.

LATE in the summer of 1956 my wife and I moved to Washington. We lived with two daughters, a cat, and a dog, in Chevy Chase; every day I would drive to work through Rock Creek Park, past the zoo. I worked across the street from the Capitol, at the Library of Congress. I knew Washington fairly well, but had never lived there; I had been in the army, but except for that had never worked for the government.

Some of the new and some of the old things there—I was often reminded of the army—had a good deal of effect on me: after a few weeks I began to write a poem. I have most of what I wrote, though the first page is gone; the earliest lines are

any color
My print, that has clung to its old colors
Through many washings; this dull null
Navy I wear to work, and wear from work, and so
~~And so to bed~~ To bed
With no complaint, no comment—neither from my chief,
nor
The Deputy Chief Assistant, ~~from~~ his chief,
Nor nor
~~From~~ Congressmen, ~~from~~ their constituents—
~~thin~~
Only I complain; this ~~poor~~ worn serviceable . . .

The woman talking is a near relation of women I was seeing there in Washington—some at close range, at the Library—and a distant relation of women I had written about before, in "The End of the Rainbow" and "Cinderella" and "Seele im Raum." She is a kind of aging machine-part. I wrote, as they say in suits, "acting as next friend"; I had for her the sympathy of an aging machine-part. (If I was also something else, that was just personal; and she also was something else.) I felt that one of these hundreds of thousands of government clerks might feel all her dresses one dress, a faded navy blue print, and that dress her body. This work- or life-uniform of hers excites neither complaint, nor comment, nor the mechanically protective *No comment* of the civil servant; excites them neither from her "chief," the Deputy Chief Assistant, nor from his, nor from any being on any level of that many-leveled machine: all the system is silent, except for her own cry, which goes unnoticed just as she herself goes unnoticed. (I had met a Deputy Chief Assistant, who saw nothing remarkable in the title.) The woman's days seem to her the going-up-to-work and coming-down-from-work of a worker; each ends in *And so to bed*, the diarist's conclusive unvarying entry in the daybook of his life.

These abruptly opening lines are full of duplications and echoes, like what they describe. And they are wrong in the way in which beginnings are wrong: either there is too much of something or it is not yet there. The lines break off with *this worn serviceable*—the words can apply either to her dress or to her body, but anything so obviously suitable to the dress must be intended for the body. *Body that no sunlight dyes, no hand suffuses*, the page written the next day goes on; then after a space there is *Dome-shadowed, withering among columns,/ Wavy upon the pools of fountains, small beside statues* . . . No sun colors, no hand suffuses with its touch, this used, still-useful body. It is subdued to the element it works in: is shadowed by the domes, grows old and small and dry among the columns, of the buildings of the capital; becomes a reflection, its material identity lost, upon the pools of the fountains of the capital; is dwarfed beside the statues of the capital—as year by year it passes among the public places of this city of space and trees and light, city sinking beneath the weight of its marble, city of graded voteless workers.

The word *small,* as it joins the reflections in the pools, the trips to the public places, brings the poem to its real place and subject—to its title, even: next there is *small and shining,* then (with the star beside it that means *use, don't lose*) *small, far-off, shining in the eyes of animals;* the woman ends at the zoo, looking so intently into its cages that she sees her own reflection in *the eyes of animals, these wild ones trapped/ As I am trapped but not, themselves, the trap . . .* The lines have written above them *The Woman at the Washington Zoo.*

The next page has the title and twelve lines:

This print, that has kept the memory of color
Alive through many cleanings; this dull null
Navy I wear to work, and wear from work, and so
To bed (with no complaints, no comment: neither
from my chief,
The Deputy Chief Assistant, nor her chief,
Nor his, nor Congressmen, nor their constituents
~~*wan*~~
—Only I complain); this ~~plain,~~ worn, serviceable
sunlight
Body that no ~~sunset~~ dyes, no hand suffuses
But, dome-shadowed, withering among columns,
Wavy beneath fountains—small, far-off, shining
~~*wild*~~
In the eyes of animals, these beings trapped
As I am trapped but not, themselves, the trap . . .

Written underneath this, in the rapid, ugly, disorganized handwriting of most of the pages, is *bars of my body burst blood breath breathing—lives aging but without knowledge of age / Waiting in their safe prisons for death, knowing not of death;* immediately this is changed into two lines, *Aging, but without knowledge of their age,/ Kept safe here, knowing not of death, for death*—and out at the side, scrawled heavily, is: *O bars of my own body, open, open!* She recognizes herself in the animals—and recognizes herself, also, in the cages.

Written across the top of this page is *2nd and 3rd alphabet.* Streets in Washington run through a one-syllable, a two-syllable, and a three-syllable (Albemarle, Brandywine, Chesapeake . . .) alphabet, so that people say about an address: "Let's see, that's in the second alphabet, isn't it?" It made me think of Kronecker's, "God made the integers, all else is the work of man"; but it seemed right for Washington to have alphabets of its own—I made up the title of a detective story, *Murder in the Second Alphabet.* The alphabets were a piece of Washington that should have fitted into the poem, but didn't; but the zoo was a whole group of pieces, a little Washington, into which the poem itself fitted.

Rock Creek Park, with its miles of heavily wooded hills and valleys, its rocky stream, is like some National Forest dropped into Washington by mistake. Many of the animals of the zoo are in unroofed cages back in its ravines. My wife and I had often visited the zoo, and now that we were living in Washington we went to it a great deal. We had made friends with a lynx that was very like our cat that had died the spring before, at the age of sixteen. We would feed the lynx pieces of liver or scraps of chicken and turkey; we fed liver, sometimes, to two enormous white timber wolves that lived at the end of one ravine. Eager for the meat, they would stand up against the bars on their hind legs, taller than a man, and stare into our eyes; they reminded me of Akela, white with age, in the *Jungle Books*, and of the wolves who fawn at the man Mowgli's brown feet in *In the Rukh*. In one of the buildings of the zoo there was a lioness with two big cubs; when the keeper came she would come over, purring her bass purr, to rub her head against the bars almost as our lynx would rub his head against the turkey-skin, in rapture, before he finally gulped it down. In the lions' building there were two black leopards; when you got close to them you saw they had not lost the spots of the ordinary leopards—were the ordinary leopards, but spotted black on black, dingy somehow.

On the way to the wolves one went by a big unroofed cage of foxes curled up asleep; on the concrete floor of the enclosure there would be scattered two or three white rats—stiff, quite untouched—that the foxes had left. (The wolves left their meat, too—big slabs of horse-meat, glazing, covered with flies.) Twice when I came to the foxes' cage there was a turkey-buzzard that had come down for the rats; startled at me, he flapped up heavily, with a rat dangling underneath. (There are usually vultures circling over the zoo; nearby, at the tennis courts of the Sheraton-Park, I used to see vultures perched on the tower of WTTG, above the court on which Defense Secretary McElroy was playing doubles—so that I would say to myself, like Peer Gynt: "Nature is witty.") As a child, coming around the bend of a country road, I had often seen a turkey-buzzard, with its black wings and naked red head, flap heavily up from the mashed body of a skunk or possum or rabbit.

A good deal of this writes itself on the next page, almost too rapidly for line-endings or punctuation: *to be and never know I am when the vulture buzzard comes for the white rat that the foxes left May he take off his black wings, the red flesh of his head, and step to me as man—a man at whose brown feet the white wolves fawn—to whose hand of power / The lioness stalks, leaving her cubs playing / and rubs her head along the bars as he strokes it.* Along the side of the page, between these lines, two or three words to a line, is written *the animals who are trapped but are not themselves the trap black leopards spots, light and darkened, hidden except to the close eyes of love, in their life-long darkness, so I in decent black, navy blue.*

As soon as the zoo came into the poem, everything else settled into it and was at home there; on this page it is plain even to the writer that all the things in the poem come out of, and are divided between, color and colorlessness. Colored women and colored animals and colored cloth—all that the woman sees as her own opposite—come into the poem to begin it. Beside the typed lines are many hurried phrases, most of them crossed out: *red and yellow as October maples rosy, blood seen through flesh in summer colors wild and easy natural leaf-yellow cloud-rose leopard-yellow, cloth from another planet the leopards look back at their wearers, hue for hue the women look back at the leopard.* And on the back of the vulture's page there is a flight of ideas, almost a daydream, coming out of these last phrases: *we have never mistaken you for the others among the legations one of a different architecture women, saris of a different color envoy impassive clear bullet-proof glass lips, through the clear glass of a rose sedan color of blood you too are represented on this earth . . .*

One often sees on the streets of Washington—fairly often sees at the zoo—what seem beings of a different species: women from the embassies of India and Pakistan, their sallow skin and black hair leopard-like, their yellow or rose or green saris exactly as one imagines the robes of Greek statues before the statues had lost their colors. It was easy for me to see the saris as cloth from another planet or satellite; I have written about a sick child who wants "a ship from some near star/ To land in the yard and beings to come out/ And think to me: 'So this is where you are!' " and about an old man who says that it is his ambition to be the pet of visitors from another planet; as an old reader of science fiction, I am used to looking at the sun red over the hills, the moon white over the ocean, and saying to my wife in a sober voice: "It's like another planet." After I had worked a little longer, the poem began as it begins now:

The saris go by me from the embassies.

Cloth from the moon. Cloth from another planet.
They look back at the leopard like the leopard.

And I . . . This print of mine, that has kept its color
Alive through so many cleanings; this dull null
Navy I wear to work, and wear from work, and so
To my bed, so to my grave, with no
Complaints, no comment: neither from my chief,
The Deputy Chief Assistant, nor his chief—
Only I complain; this serviceable
Body that no sunlight dyes, no hand suffuses
But, dome-shadowed, withering among columns,
Wavy beneath fountains—small, far-off, shining
In the eyes of animals, these beings trapped
As I am trapped but not, themselves, the trap,
Aging, but without knowledge of their age,
Kept safe here, knowing not of death, for death
—Oh, bars of my own body, open, open!

It is almost as if, once all the materials of the poem were there, the middle and end of the poem made themselves, as the beginning seemed to make itself. After the imperative *open, open!* there is a space, and the middle of the poem begins evenly—since her despair is beyond expression—in a statement of accomplished fact: *The world goes by my cage and never sees me*. Inside the mechanical official cage of her life, her body, she lives invisibly; no one feeds this animal, reads out its name, pokes a stick through the bars at it—the cage is empty. She feels that she is even worse off than the other animals of the zoo: they are still wild animals—

since they do not know how to change into domesticated animals, beings that are their own cages —and they are surrounded by a world that does not know how to surrender them, still thinks them part of itself. This natural world comes through or over the bars of the cages, on its continual visits to those within: to those who are not machine-parts, convicts behind the bars of their penitentiary, but wild animals—the free beasts come to their imprisoned brothers and never know that they are not also free. Written on the back of one page, crossed out, is *Come still, you free;* on the next page this becomes

The world goes by my cage and never sees me.
And there come not to me, as come to these,
The wild ~~ones~~ beasts, sparrows pecking the llamas' grain,
Pigeons ~~fluttering to~~ settling on the bears' bread, turkey-buzzards
~~*Coming with grace first,*~~*,* ~~*then with horror*~~ ~~*Vulture seizing*~~
Tearing the meat the flies have clouded . . .

In saying mournfully that the wild animals do not come to her as they come to the animals of the zoo, she is wishing for their human equivalent to come to her. But she is right in believing that she has become her own cage—she has changed so much, in her manless, childless, fleshless existence, that her longing wish has inside it an increasing repugnance and horror: the innocent sparrows *pecking* the llamas' grain become larger in the pigeons *settling on* (not *fluttering to*) the bears' bread; and these grow larger and larger, come (with grace first, far off in the sky, but at last with horror) as turkey-buzzards seizing, no, *tearing* the meat the flies have clouded. She herself is that stale left-over flesh, nauseating just as what comes to it is horrible and nauseating. The series *pecking, settling on*, and *tearing* has inside it a sexual metaphor: the stale flesh that no one would have is taken at last by the turkey-buzzard with his naked red neck and head.

Her own life is so terrible to her that, to change, she is willing to accept even this, changing it as best she can. She says: *Vulture* [it is a euphemism that gives him distance and solemnity], *when you come for the white rat that the foxes left* [to her the rat is so plainly herself that she does not need to say so; the small, white, untouched thing is more accurately what she is than was the clouded meat—but, also, it is euphemistic, more nearly bearable], *take off the red helmet of your head* [the bestiality, the obscene sexuality of the flesh-eating death-bird is really—she hopes or pretends or desperately is sure—merely external, *clothes*, an intentionally-frightening war-garment like a Greek or Roman helmet], *the black wings that have shadowed me* [she feels that their inhuman colorless darkness has always, like the domes of the inhuman city, shadowed her; the wings are like a black parody of the wings the Swan Brothers wear in the fairy tale, just as the whole costume is like that of the Frog Prince or the other beast-princes of the stories] *and step* [as a human being, not fly as an animal] *to me as* [what you really are under the disguising clothing of red flesh and black feathers] *man*—not the machine-part, the domesticated animal that is its own cage, but man as he was first, still must be, is: the animals' natural lord,

The wild brother at whose feet the white wolves fawn,
To whose hand of power the great lioness
Stalks, purring . . .

And she ends the poem when she says to him:

You know what I was,
You see what I am: change me, change me!

Here is the whole poem:

THE WOMAN AT THE WASHINGTON ZOO

The saris go by me from the embassies.

Cloth from the moon. Cloth from another planet.
They look back at the leopard like the leopard.

And I . . .
This print of mine, that has kept its color
Alive through so many cleanings; this dull null
Navy I wear to work, and wear from work, and so
To my bed, so to my grave, with no
Complaints, no comment: neither from my chief,
The Deputy Chief Assistant, nor his chief—
Only I complain; this serviceable
Body that no sunlight dyes, no hand suffuses
But, dome-shadowed, withering among columns,
Wavy beneath fountains—small, far-off, shining

In the eyes of animals, these beings trapped
As I am trapped but not, themselves, the trap,
Aging, but without knowledge of their age,
Kept safe here, knowing not of death, for death
—Oh, bars of my own body, open, open!

The world goes by my cage and never sees me.
And there come not to me, as come to these,
The wild beasts, sparrows pecking the llamas' grain,
Pigeons settling on the bears' bread, buzzards
Tearing the meat the flies have clouded . . .
Vulture,
When you come for the white rat that the foxes left,
Take off the red helmet of your head, the black
Wings that have shadowed me, and step to me as man,
The wild brother at whose feet the white wolves fawn,
To whose hand of power the great lioness
Stalks, purring . . .
You know what I was,
You see what I am: change me, change me!

Malraux and the Statues at Bamberg

IT IS no use to tell you to read *The Voices of Silence:* if you care for art and know how to read, you have read it or will read it. And if you don't care for art but know about it instead, and have spent your life stopping up the holes in your dressing-gown with the canvases of the universe—even then you will read it, so as to be able to call Malraux a phrase-making amateur standing on the shoulders of better art-historians. And if you say this, there will be something in what you say. Malraux does stand head and shoulders above most writers on art, though I doubt that he gets this way simply by basing himself on them; he is an amateur—he speaks with all of the exaggeration of love and some of its errors; and he makes phrases as naturally as—or rather, considerably more naturally than—he breathes. (His last breath is going to be drawn in the Pantheon, under the subjugated eyes of Academicians, and his earlier breaths have been changed by knowing this.)

Malraux's book is a long, lyrical, aphoristic, oratorical, wonderfully illustrated Discourse on the Arts of this Earth, with space for Celtic coins, Van Meegeren's Vermeers, any artist who ever was, fairy tales, religions, a history of taste, the drawings of the insane, best-sellers, the influence of Tintoretto on cameramen: it is a kind of (very elevated) Flea-Market of the Absolute, with room even for a remark about paintings at the Flea-Market. Malraux's intelligence, imagination, and originality manifest themselves as much in his choice of subjects as in what he has to say. His work is not art history, exactly, but a kind of free fantasia on themes from the history of art—still, a successful enough confusion of genres is a new genre, and Malraux's book, which now stands solitary as *Alice*, will probably have some dreadful descendants.

It is certainly one of the most interesting books ever written: Malraux writes a passage of ordinary exposition so that we breathe irregularly and jerk our heads from side to side, like spectators at a tennis match. He conducts an argument (and he can't even tell you that Art isn't Photography without having a fearful and dazzling argument that leaves you sorry you ever thought it was) as other people conduct a campaign, and his pages are full of speeches to the soldiers, of epigrams and aphorisms and passages of more-than-Tyrian purple, of *Te Deums*, of straw-men with their bowels all over the countryside: it is as if we were getting to see a *Massacre of the Innocents* begun by Uccello, concluded by Caravaggio, and preserved for us, I do not need to say miraculously, in an armory of the Knights of Malta.

Malraux tells us hundreds of times that the artist "masters" the world, "conquers" his material, "destroys" the works of the predecessors he admires; art is a "victory" over reality, the work of art "subjugates" its spectator: the root metaphor underlying Malraux's view of art is one of conquest, of victory, power, domination. (One feels: how all the animals do order each other around!) Chardin's "seeming humility" is said to involve the model's "destruction"; Malraux says in a typical sentence: "That eternal *youngness* of mornings in the Ile-de-France and that shimmer—like the long, murmurous cadences of the *Odyssey*—in the Provencal air cannot be imitated; they must be conquered." Conquered! The sentence is worthy of Cortez, of Pizarro, of those lunatics who used to think themselves Napoleon. How do you "conquer" a shimmer in the air, the youngness of morning? We do not know how to go about it; most of us do not even know how to want to go about it. Such things cannot be imitated, Malraux is right; they must be translated. The artist translates, finds an equivalent for, makes a painting or poem or piece of music that has, that shimmer—and, often, in the purity of separation, the youngness seems to us younger, the shimmer more shimmering. Art matches the world idiom for idiom; the work of art is not a conquest of the things of the world but their apotheosis. Really Malraux knows this: as he says, the buffalo in some Cro-Magnon cave is not a different buffalo but *more* buffalo—the buffalo of the world is not subjugated beneath it, conquered by it, but realized in it.*

This man who writes so wonderfully of Piero della Francesca, Latour, Vermeer, of men who would have thought the word *conquest* a profanation, cannot think of the natural world except as raw material ready for the processing that is our victory and its defeat. For Malraux, at bottom, the world is a war; and anyone who writes about *The Voices of Silence*, responding to this, will tend to make his praise warm and general and his blame cold and specific. The book is, some of the time, a marvellous evocation and appreciation of the works of art it reproduces; the rest of the time it is an argument, a fight, and we fight back. We are willing to forgive most things to a writer who cares this much for painting and sculpture, and catches us up in his caring; but the life and manners of this truly vivacious book are too contagious, we forget all about forgiving, make what speeches we can to what troops we have, gnaw our moustaches, and get out our French dictionaries so as to denounce Malraux in the style to which he is accustomed.

Malraux's passion, violence, energy seem as

* What I had remembered Malraux as saying is quite different from what he actually says: "If the Magdalenian bison is more than a sign and also more than a piece of illusionist realism—if, in short, it is a bison *other than the real bison*—is this merely due to chance?"

genuine as they are habitual, though the form in which they are expressed and the emotions by which they are accompanied are often stock. I once visited a pottery where all the dogs and cats were named after characters in the *Ring*, and where all the shapes of the pots, as the potter told me, "had authority." All Malraux's sentences "have authority," but only his more timid or less informed readers are likely to remain submissive beneath it: Malraux writes in a language in which there is no way to say "perhaps" or "I don't know," so that after a while we grow accustomed to saying it for him. But we need to say it less often than would be supposed: how many men have written about art with better taste, with more intelligence, with a keener sympathy, with a more extraordinary scope and grasp and intensity, and have alloyed these with a rhetoric so grandiose, sentiments so conventionally theatrical, and an obsession with power so radical, that a book of theirs can seem to us a miracle which we partly dislike?

When Malraux wants to assert a proposition about which he, alone among mortals, feels no doubts, he prefaces it with—and I combine two of his favorite introductory remarks—"WHO CAN FAIL TO SEE THAT, *regardless of what everybody says . . .*" But ordinarily he proves his propositions. He says, for instance, that "the ill-success of *The Night Watch* was inevitable," and goes on to show us why it was inevitable. He is completely successful: when we have finished his paragraphs we can see that the painting *had* to be disliked. If we happen to have learned that as a matter of fact it was greatly liked, and that its failure is a poetically just myth, we are troubled to see Malraux's method so powerful. The connections of European art with Christianity are more enlightening, if less surprising, than its connections with double-entry bookkeeping, so that Malraux's semi-religious determinism is a good deal better than the economic determinism which tells us that Masaccio's outlines are as firm as they are because the financial position of the rising middle class was as sound as it was. But both methods have the same fault: they are too powerful. By using either we can show just why everything necessarily was what we already know it to have been—and we can often, in the process, distort (or neglect to see closely enough or disinterestedly enough) what everything was.

We say with a sigh: "The ways of God are inscrutable." To the critic of art the ways of art are inevitable, and he explains with a smile why everything had to happen as it did. (He can explain it only after it happens, not before—but, everything has its limitations. In everyday life, a crude sphere, we call someone whose explanations have these limitations a Monday morning quarterback.) The critic of art often does all that he can to make the ways of art inevitable, saying without any smile: "Certainly no representational painting of the first importance could be produced *now;* certainly no diatonic composition of the first importance could be produced *now*." Schönberg said that there were a great many good pieces still to be composed in the key of C Major, and his sentence is as inspiring to me, as a human being, as is Cromwell's: "I beseech you, in the bowels of Christ, believe that you may be mistaken!"

What I am saying is very obvious, and if anything is obvious enough it seems almost to give us the right to ignore it. Analysts of society or art regularly neglect what is, for the parts of it their explanation is able to take account of, and then go on the assumption that their explanation is all that there is. (If the methods of some discipline deal only with, say, what is quantitatively measurable, and something is not quantitatively measurable, then the thing does not exist for that discipline—after a while the lower right hand corner of the inscription gets broken off, and it reads *does not exist.*) But if someone has a good enough eye for an explanation he finally sees nothing inexplicable, and can begin every sentence with that phrase dearest to all who professionally understand: *It is no accident that* . . . We should love explanations well, but the truth better; and often the truth is that there *is* no explanation, that so far as we know it is an accident that . . . The motto of the city of Hamburg is: *Navigare necesse est, vivere no necesse.* A critic might say to himself: for me to know *what* the work of art is, is necessary; for me to explain *why* it is what it is, is not always necessary nor always possible.

Let me illustrate all this by examining Malraux's treatment of six statues, three at Rheims and three at Bamberg. He writes that at Rheims the art of antiquity, of the smile, of "smoothly modeled planes, of supple garments and gestures," was at last able to be resuscitated, in order "to voice the concord between man and what transcends him, the last act of the Incarnation." He reminds the reader that "there are classical precedents for the way the Master of the Visitation treated drapery," persuades him (unnecessarily, but in sensitive and beautiful detail) that the Virgin *is* Gothic, not classical, and finishes by explaining why the sculptor was able to make the Mary and Elizabeth of this *Visitation* what they are: "When man had made his peace with God and once again order reigned in the world, the sculptors found in the art of antiquity a means of expression ready to hand. If we turn east to Bamberg, where this reconciliation was less complete, we find that its Virgin gives an impression of being much earlier than the Rheims Virgin, from which, nevertheless, it derived. Gazing with eyes still misted with fears of hell, above that miraculously apt fracture which makes her face the very effigy of Gothic death, the St. Elizabeth of Bamberg seems to contemplate her 'prototype' of Rheims across an abyss of time."

This last sentence may itself seem to us a work of art; certainly we seem to ourselves to feel more, to understand better, now that the weights and relations of these things have been shown to us, understood for us. But if we look at the Bamberg and Rheims *Visitations*, the Bamberg Rider and its "prototype" at Rheims, and read what is known about the circumstances of their making, we find that Malraux has understood for us too swiftly and too well: some of his facts are not facts at all, and —what matters far more—some of his feeling for, his seeing of, these statues has been distorted by his understanding of them, by the thesis to which the statues have been required to testify.

Certainly Bamberg was influenced by Rheims; but the Bamberg *Visitation* often is dated before the Rheims *Visitation*—which *was* the prototype? That the Rheims Virgin and St. Elizabeth look just as they look doesn't puzzle Malraux, but it puzzles everyone else: sober art-historians sound like writers of detective-stories as they try to account for it. Morey says, for instance: "It is not impossible that a German sculptor, schooled in the early atelier of Rheims, returned to work at Bamberg, and later joined the eclectic group which finished the façade of Rheims cathedral. This would explain the Teutonic Mary and Elizabeth of the *Visitation* at Rheims, and the general identity of style of the Elizabeth in this group with the Elizabeth of another *Visitation* in the ambulatory at Bamberg." It would explain it—unless we look at the statues. The three Bamberg statues which have "prototypes" at Rheims—the Virgin, the St. Elizabeth, and the Rider—are a family of masterpieces, as like as sister, mother, and brother; the Rheims Virgin,

St. Elizabeth, and Philip Augustus vary from sublimity to pure commonplace. Philip (purse-lipped, tremulous-lidded, his face closed uncertainly, almost pedantically, about the cares of power) looks as though he were dreaming that he is the other, that Rider who—profound, ambiguous, weighing, full of a strength touching in its delicacy and forbearance, of an innate, almost awkward elegance—looks out with eyes more considering, more deeply set, more widely set than other eyes, so that we see in him one of the great expressions of man's possibility, of that grace which comes upon him too naturally for him to be aware either of its source or of its presence.

The Rheims St. Elizabeth is, as Malraux would say, a Stoic's mother that has somehow found a soul, a quiet, sad, beautiful statue which we forget as we look at the Bamberg St. Elizabeth. *Her* eyes are misted neither by "fears of hell" nor by anything else—here Malraux's taste, his mere ability to see, have been debauched by his theory—but are calmer and less changing than the stone in which they are carved. Man's ability to bear and disregard—to look out into, to look out past, anything in his world—has been expressed as well in a few other works of art, but never, I think, better: she seems to look out into that Being which has cancelled out the Becoming which the Rider looks into and is.

But the two Virgins are most puzzling of all, if we want to *explain* them: how can one masterpiece be derived from another that contradicts it? The

Virgin of Rheims' humanity, which Malraux describes so beautifully, is too wonderful for us to be willing to compare her unfavorably with anything, yet the Virgin of Bamberg's inhumanity (if a Fate can be called inhuman) is almost as wonderful: her reserved, brooding, slightly too full young face, with the future held uneasily in its fullness, is like the first premonition of one of Michelangelo's sibyls; as we look at the curve of this body at once extended to us and withdrawn from us, at the grave, dimpled, half-archaic smile of a troubling and ununderstandable benediction, we feel more than ever the inscrutability of God's ways and our own.

Do the Rheims figures look as they look, do the Bamberg figures look as *they* look, because at one place the "reconciliation" of God and man was "less complete"? While we read Malraux, we understand; while we look at the statues we do not understand, but we are looking at the statues.

And, much of the time, Malraux is looking with us; but a historian, a critic, cannot always stand idly looking and feeling, but must explain things. While he explains them gently and consideringly, with tact and insight and forbearance—explains them partially, only partially—*then* we may see them as we have never seen them before, but as he has. Often Malraux does this—with Latour, for instance—but often he is a rough explainer. A little of Rheims fits into his ideological scheme perfectly, as the climax of Gothic art; the specific qualities of Bamberg and Naumburg do not, so that in spite of liking them so much that he refers to them again and again, he "filters" them out of his main argument, and regards them as belated survivals of an earlier stage of Gothic development. Similarly, he says about Vermeer (and he writes with a wonderful feeling for the depth and poetry of a painter so often called "limited" or a "jeweler") that "the depiction of a world devoid of value can be magnificently justified by an artist who treats *painting itself* as the supreme value." He goes on to say about his favorite Vermeer, *The Love Letter:* "The letter has no importance, and the woman none. Nor has the world in which letters are delivered; all has been transmuted into painting." If the picture had been called *The Visitation,* how willingly Malraux would have accepted the importance of the visit, of the woman, and of the world in which visits are made! (One feels like saying: Vermeer's canvases are as full of values as is Spinoza's *Ethics;* they might even be used as illustrations for it.) But Malraux is quite sensitive to religious and quasi-religious values, quite insensitive to others; the sciences, for him, are not much more than what has produced television-sets and the atomic bomb. The quieter personal and domestic values—what St. Jerome felt for his lion instead of what he felt for his church —hardly exist for Malraux; his mind is large and public. Someone said that in the ideal dictatorship everything would be either forbidden or obligatory; in Malraux's world everything is either heroic, ignoble, or irrelevant. The world itself, for Malraux, has become a kind of rhetoric. Any religion attracts him as a work of art, a source of style, an incarnation of values, and yet you can hardly imagine his believing in one—in *one*. You can say of him what you can say of any true rhetorician: that rhetoric frees us from all claims except its own.

The Voices of Silence, if it had a denotative instead of a connotative title, might be called *Religious, Anti-Religious, and Proto-Religious Art.* Malraux detests "the incapacity of modern civilization for giving form to its spiritual values," and says in a beautiful and heartfelt sentence: "On the whole face of the globe the civilization that has conquered it has failed to build a temple or a tomb." He goes on: "Agnosticism is no new thing: what is new is an agnostic culture. Whether Cesare Borgia believed in God or not, he reverently bore the sacred relics, and, while he was blaspheming among his boon companions, St. Peter's was being built." Whether he believes in God or not, Malraux passionately believes in the necessity of a culture's believing, and he reverently bears into the only church any longer possible to us, the Museum without Walls, the sacred relics of the religious arts of the past.

Malraux feels that European painting first embodied religious values, next embodied poetic and humanistic values, and finally rejected "*all* values that are not purely those of painting." Just as religious art distorted Nature into forms that would express the values of religion, so modern painting distorts it into forms that will express the values of painting: "the quality modern art has in common with the sacred arts," Malraux writes, "is not that, like them, it has any transcendental significance, but that, like them, it sponsors only such forms as are discrepant from visual experience . . . Our style is based on a conviction that the only world which matters is other than the world of appearances." One might suppose that in the long run "the world of appearances" would have to matter to that creator of a new world of appearances the painter. But Malraux despises the arts of "illusionist realism," of "delectation." "When I hear the word Nature," Malraux might say, "I reach for my revolver." In his scheme of things there are earlier artists and their schemata, the new artist and his schema, and far back in the corner under a dunce's stool, cowering dully, a dwarfed Mongoloid, Nature. It is almost out of the picture; and yet Malraux is uneasy at having it there, he wants it all the way out.

The artist and Nature, as Malraux conceives them, are almost exactly like Henry James and that innocent bystander who gives James the germ of one of his stories. "Just outside Rye the other day," the man begins, "I met a—" "Stop! stop!" cries James. "Not a word more or you'll spoil it!" Most of the European artists Malraux writes about had a neurotic, life-long compulsion to *look at things:* they made studies, made sketches, made dissections, paid models, hired maids because of the way the maids' skin took the light; some of them lived half their lives in the middle of the landscape, getting red and wrinkled with its suns, getting pneumonia from its rains—so that an unsympathetic observer can say of their style what Malraux says of Corot's, that it "tells of a long conflict with Nature (which he was apt to confound with the pleasure of visits to the country)." Nature had their deluded, heartfelt tributes, their "assertions that they were her faithful servants." "Certain masters," Malraux observes, "even claimed that this submission contributed to their talent." Some loved to say that Nature had been their master.

When they say this Malraux of course realizes that this is not what they meant; he knows that he wouldn't let a thing like Nature be *his* master. (Actually Malraux wouldn't let anything be his master: his motto is *Vici, vici, vici.*) He explains what they did mean: "When Goya mentioned Nature as being one of his three masters he obviously meant, 'Details I have observed supply their accents to ensembles I conjure up in my imagination.'" Obviously. "When Delacroix spoke of Nature as a dictionary he meant that her elements were incoherent." What they said either meant something else, or they didn't mean what they said, or they meant what they said but just didn't know: "No great painter," Malraux concludes, "has ever talked as we would like him to talk." When van Gogh made a painting of a chair that almost pushes the chair in your face, that says like a child: *Look, there's my chair!*—at that moment, Malraux writes, "the conflict between the artist and the outside world, after smoldering for so long, had flared up at last." Malraux singularly, uniquely, inimitably sees that in this picture van Gogh had declared war on the chair. But what puzzles Malraux is: why did he, why did they all, wait so long to declare it? If only—

"Chardin's *Housewife*," writes Malraux, "might be a first-class Braque dressed-up enough to take in the spectator." "Corot," writes Malraux, "makes of the landscape a radiant still life; his *Narni Bridge*, *Lake of Garda*, and *Woman in Pink* are, like the *Housewife*, dressed-up Braques." If only Chardin and Corot had undressed their Braques! If only Chardin had got rid of that housewife, Corot of that landscape, the world would have had two more Braques, 1760 and 1840 would have been 1920, and Malraux wouldn't be having to persuade us to pay no attention to what Chardin and Corot said about Nature. (Somehow it never seems to occur to him to call Piero della Francesca's *Resurrection* a dressed-up Cézanne, and to imply that if Piero had only got rid of Christ and those soldiers he would have had a first-class Cézanne: religion matters.) Malraux spends so much time persuading his readers that art isn't a photographic imitation of Nature that it gives portions of his book a curiously old-fashioned look—he needs to persuade most of his readers, today, that art has *any* relation to Nature.

Of the quasi-aesthetic organization of visual perception itself—an organization that is at the root of aesthetic organization—Malraux is ignorant; for him Koffka, Köhler, Gombrich, and the rest might never have existed. He understands "the world of appearances," our perception of Nature, only as a kind of literal photograph of formless raw material—raw material that demands the immediate transformation, the supernatural distortion, of a religious, transcendental art. (An old sexual metaphor underlies Malraux's view of aesthetic creation: a masculine Art forms and conquers a helpless, formless, feminine Nature.) Malraux is willing to accept a modern art that no longer embodies religious values, if it rejects "the world of appearances," is "discrepant from visual experience." Yet his acceptance is the provisional acceptance we give to something that is only a transitional stage; "akin to all styles that express the transcendental and unlike all others, our style seems to belong to some religion of which it is unaware," "is nearing its end," "cannot survive its victory intact." Our culture "will certainly transform modern art"; after having "conquered and annexed" the religious and transcendental arts of the past, it will see its own art become a new religious and transcendental art of a nature we cannot now fathom: "whether we desire it or not, Western man will light his path only, by the torch he carries, even if it burns his hands, and what that torch is seeking to throw light on is everything that can enhance the power of man." (The disadvantages of so oratorical a style of understanding as Malraux's are terribly apparent in the last sentence.) Malraux adores power, and is willing to accept the understanding that sometimes goes along with power, if that power and that understanding are personal, aesthetic, religious; when they are the impersonal power and understanding of science, that conquers in the long run by being submissive and observant in the short run, by first imagining and then seeing whether the fact fits, Malraux has no interest in them.

Reality is what we want it to be or what we do not want it to be, but it is not our wanting or our not wanting that makes it so. Yet with sufficient mastery the critic can have reality almost what he wants it—and Malraux's temperament is very masterful. (The last time I read *The Voices of Silence* I thought longingly of the submissiveness of Sir Kenneth Clark's *The Nude*—just as, the last time I heard *Parsifal*, I couldn't help thinking of *Falstaff*.) The soldier Descartes philosophized while at war, in quiet winter quarters; Malraux has to philosophize in the midst of a war which he himself is staging, a war that rages through every season. That Malraux's is a book of "tremendous philosophical and moral importance"—Edmund Wilson says so—I cannot believe. How could such a book have the *facility*—of thought, feeling, and expression—that this book so often has? Yet it is a book that shows better than any but a few others what art has been to man. Often Malraux writes as well about some painter or sculptor, some style or influence, some metamorphosis in taste, as anyone I have read, and his intelligence, his dramatic imagination, his passionate absorption, the sheer liveness of his experience and knowledge are extraordinary; if he is an intermittent trial, he is a continual delight. His virtues are so dazzlingly apparent that one writes at exaggerated length about his faults; but his book is a work of art, and we judge it as we judge a work of art: by its strengths.

What is worst about it, as a work of art, is the way in which some of it is written. It has thousands of arresting or moving or conclusive or exactly realizing sentences, and hundreds of sentences that are coarsely, theatrically, and conventionally rhetorical. One cannot do justice to Malraux's good writing, since there is so much of it, but one can certainly do justice to the bad. "The Oriental night of blood and doom-fraught stars" is the phrase he finds for Byzantium; he calls the world of primitive artists "a nether world of blood, and fate-fraught stars"; he writes that "neither blood, nor the dark lures of the underworld, nor the menaces of doom-fraught stars have at all times prevailed against that soaring hope which enabled human inspiration, winged with love, to confront the palpitating vastness of the nebulae with the puny yet indomitable forms of Galilean fishermen or the shepherds of Arcadia." Who would have thought those old stars had so much doom in them? Malraux says about the Dutch: "We tend to overlook that glorious page of Dutch history, and even today you will hear people talking, as of quaint figures in picture-postcards, of a nation that put up a stout resistance to Hitler's hordes, and has led the world in postwar reconstruction." This is the *lingua-franca* of vice-presidents and major-generals, the tongue in which the Dean talks to the *pompier*, and needs no translator. Malraux even ends his book, his beautiful book, with this sentence: "And that hand whose waverings in the gloom are watched by ages immemorial is vibrant with one of the loftiest of the secret yet compelling testimonies to the power and glory of being Man." Imagine—I won't say Rilke's—imagine Degas' face as he read such a sentence!

As for another author, the man who wrote *La Condition Humaine;* the man who writes about the first "retrograde" art: "Thereafter Byzantium reigned alone. The age which was discovering the sublimity of tears showed not a weeping face," and who goes on: "As much genius was needed to obliterate man at Byzantium as to discover him on the Acropolis"; the man who tells how the yielding feminine smile of the Greek statues was changed by the Asiatic sculptors into "something sterner, hewn in the cliffside: the lonely smile of the men of silence"; who says of Botticelli's figures that "knots of fine-spun lines enwrap their shining smoothness"; who writes about the end of the Middle Ages and the beginning of the Renaissance: "To restore to life that slumbering populace of ancient statues, all that was required was the dawn of the first smile upon the first mediaeval figure," ends his chapter there, and begins the next: "How very timid was that smile!"—as for the man who wrote these sentences, how does *he* feel about the man who wrote the others? We are almost willing to use the terms of Malraux's effective and misleading distinction between Michelangelo and Signor Buonarroti, Paul Cézanne and Monsieur Cézanne, and to say that it is André Malraux who is responsible for the grandeur of some of these sentences, and Monsieur Malraux, the well-known politician and man of action, who is responsible for the vulgar grandiosity of the others. But if we said so we should be making Malraux's mistake: it is the same man who is responsible for both, and it is our task to understand how this is possible. When we read what Goethe says about men we are ashamed of what we have said; when we read what he says about paintings and statues we are ashamed of what Goethe has said. It is one of the merits of Malraux's book that it shows, perhaps more forcibly and vivaciously than any other, why it is historically possible for us to feel this; or to feel as we feel when we read, in Berenson, that Uccello "in his zeal forgot local color—he loved to paint his horses pink and green—forgot action, forgot composition, and, it need scarcely be added, significance." A few pages later we read about a painter, more to be despised than pitied, who made "the great refusal," and who degenerated until "at his worst he hardly surpasses the elder Breughel." I suppose I ought to say that it is only *our* taste these judgments appal, that after a while the wheel will have come full circle, and another age will smile at Malraux's judgments of Uccello and Breughel, nod approvingly at Berenson's . . . Well, unhappy the age that does so! May it be further accursed!

I have talked of the faults or exaggerations of *The Voices of Silence* far too much for justice, so let me say that I have worn into sections the unbound copy I first owned and wrote about, and cannot look at my bound copy without a surge of warmth and delight: if I knew a monk I would get him to illuminate it. Who has ever picked illustrations like Malraux? He has worked hard rewriting the old version of the book: has changed, added, omitted, rearranged, more than one would have thought possible; since the phrasing has been improved and since the blank spaces that gave the old version a somewhat disjointed, aphoristic look have been done away with, it now has more of a Spenglerian weight and continuity, and less of the I-am-just-thinking-for-you air that the first version occasionally had—and had attractively, I thought. Malraux still thinks that famous Scythian deer, antlered from nose to tail, a horse, and prints it opposite Degas' for comparison; and he still believes that Schumann composed to the smell of rotten apples, but that you can't smell them in his music. Poor Schiller! I'd as soon see Eve, Newton, and Gregor Samsa deprived of their apples.

A Wilderness of Ladies

BUCK DUKE says to Mamma, as he brings her the milk-pail full of wild plums:

> *"Sour! Your eyes'll water, Miss Tempe!*
> *But sweet, too."*

The taste of someone else's life—and while you are reading the poems of Eleanor Taylor you are someone else—is almost too sour to be borne; but sweet, too. The life is that of one woman, one (as the census would say) housewife; but a family and section and century are part of it, so that the poems have the "weight and power,/ Power growing under weight" of a world. Some of this world is grotesquely and matter-of-factly funny, some of it is tragic or insanely awful—unbearable, one would say, except that it is being borne. But all of it is *so*, seen as no one else could see it, told as no one else could tell it.

The poems and poet come out of the Puritan South. This Scotch Presbyterianism translated into the wilderness is, for her, only the fierce shell of its old self, but it is as forbidding and compulsive as ever: the spirit still makes its unforgiving demands on a flesh that is already too weak to have much chance in the struggle. The things of this world are "what Ma called poison lilies, sprouting/ From Back Bunn's meadow resurrectionwise/ But with a sinful pink stain at the throat." So much, still, is sin! Blaming the declining West, a character in one of the poems says hotly: "You talk so much of rights, now;/ You ask so seldom what your duties are!" The poet knows too well for asking what her duties are, and has no rights except the right to do right and resent it: her "Lord, help me to be more humble in this world!" is followed without a pause by the exultant "In that Great, Getting-Up Morning, there will be another song!" She cannot permit herself—the whole life she has inherited will not permit her—to be happy, innocently bad, free of these endless demands, this continual self-condemnation. Frost speaks of a world where work is "play for mortal stakes," but here everything is work for mortal stakes, and harder because of the memory of play, now that nothing is play. (I once heard a woman say about buying new clothes for a trip to Europe: "It's work, Mary, it's work!"—a very Protestant and very ethical sentence.)

First there were her own family's demands on the girl, and now there are the second family's demands on the woman; and worst of all, hardest of all, are the woman's demands on herself—so that sometimes she longs to be able to return to the demands of the first family, when the immediate world was at least childish and natural, and one still had child allies in the war against the grown-ups. Now the family inside—the conscience, the superego—is a separate, condemning self from which there is no escape except in suicide or fantasies of suicide, the dark rushing not-I into which the I vanishes. And which, really, is the I? The demanding conscience, or the part that tries to meet—tries, even, to escape from—its demands? In one poem the chain gang guard envies the prisoner who still needs a guard, who cannot escape because of the rifle outside, the guard outside; the guard himself no longer has to be guarded, says in despairing mockery: "Here I stand! loaded gun across me—/ As if I'd get away!" The world is a cage for women, and inside it the woman is her own cage. In longing regression, this divided self—"riding the trolley homeward this afternoon/ With the errands in my lap"—would willingly have let it all slide from her lap, would willingly have "disfestooned my world—/ A husband, more or less!/ A family, more or less!/ To have alighted to a cup of kettle-tea/ And someone/ To whom I could lie merrily,/ Use malapropisms, be out-of-taste"; to the sister who is "more than one-flesh-and-blood,/ Almost another I." The self, two now, longs for that first world in which it and another were, almost, one; longs to return to the make-believe tea that preceded the real tea of the grown-ups, the tea that, drunk, makes one a wife and mother.

The world of the poems is as dualistic as that of Freud; everything splits, necessarily, into two warring opposites. This fault along which life divides, along which the earthquakes of existence occur, is for the poet primal. It underlies all the gaps, disparities, cleavages, discontinuities that run right through her; she could say with Emerson: "There is a crack in everything that God has made." She says about her sister and herself: "The wars of marriage and the family burst around us"; but these are only external duplicates of the war inside, the war of self and self. Life is a state of siege, of desperate measures, forlorn hopes, last extremities—is war to the last woman. Carried far enough, everything reduces to a desperate absurdity; one can say about the poems themselves what one of the poems says about a man: "You were a mortal sheen/ Flickering from the negative." The poems' Religious Wars, wars of conscience, go over into wars of anxiety and anguish, of neurosis or psychosis: "For me the expected step sinks,/ The expected light winks/ Out . . ." The water of sexuality, of unknown experience, that the child shrank from and that the woman longs to drown in, freezes into glass, gems, the hard "stock-dead" fixity of catatonia:

Oh to have turned at the landing
And never have sounded the bell
That somehow thrust me into this room
Beaded with eyes, painfully held
To the liable frame I illume.
Could life stop, or go on!

But olives dangle crystal stems,
And that clock muffles its French tick
In those elaborate kiss-me-knots;
Does it, too, hate its gems?

In frigid aberration, she spoils the life she ought to nourish: "Each year I dug and moved the peonies/ Longing to flare/ Fat and chemically by the well-slab,/ Ingrown./ Each day I opened the drawer and/ Scanned the knives." And the warped spirit (after it has desperately demanded, from outside, the miracle that alone could save it: "You should have struck a light/ In the dark I was, and/ Said, Read, Be—be over!") ends in an awful negative apotheosis, as it cries: "Not in the day time, not in the dark time/ Will my voice cut and my poison puff/ My treasures of flesh,/ My gems of flashing translucent spirit,/ Nor my caress shatter them." When in another poem a patient says: "It nettled me to have them touch my dog/ And say in their dispelling voices, *Dog*," the helpless, fretful, loving-its-own-psychosis voice of the psychotic is so human that your heart goes out to it, and can neither pigeonhole it nor explain it away.

The violent emotion of so much of the poetry would be intolerable except for the calm matter-of-factness, the seriousness and plain truth, of so much else; and except for the fact that this despairing extremity is resisted by her, forced from her, instead of being exaggerated for effect, depended on as rhetoric, welcomed for its own sake, as it is in that existential, beatnik Grand Guignol that is endemic in our age. And, too, there is so much that is funny or touching, there are so many of the homely, natural beauties known only to someone "who used to notice such things." How much of the old America is alive in lines like "She took a galloping consumption/ After she let the baby catch on fire . . ./ And Cousin Mazeppa took laudanum./ 'Why did you do it, Zeppie girl?/ Wa'n't Daddy good to you?'/ 'Pray, let me sleep!' " Byron and Liszt and Modjeska end at this country crossroads, in a name. How could old-maidly, maiden-ladyish refinement be embodied more succinctly, funnily, and finally than in:

Miss Bine taught one to violet the wrists.
"I accuse you, Mr. Stapleton,
Of excess temperance—ha ha!"
"Miss Tempe . . . I beg . . . Allow me to insist—!"

The old woman, dying among images, sees out on the dark river of death, past Cluster Rocks, "an old lady her uncles rowed across—/ The boat beneath her slipped the bank/ Just as she stepped ashore./ 'Stretched me arightsmart,' she chirped./ O to think of that dying!/ O unworthy 'Stretchedmearightsmart'!/ She glared at hell through tears." This country humor, which comes out of a natural knowledge like that of Hardy or Faulkner, can change into a gallows-humor that once or twice has the exact sound and feel of Corbière: the suicide on her way into the water mutters, "It's no good God's whistling, 'Come back, Fido' "; the cold benighted lovers flee down the blocked-off "last bat-out-of-hell roads:/ *Closed, Under Destruction.*" But these and other humors—the humors of dreams, of neurosis, of prosaic actuality—all come together in a kind of personal, reckless charm, an absolute individuality, that make one remember Goethe's "In every artist there is a germ of recklessness without which talent is inconceivable."

In the beginning there were no ladies in the wilderness, only squaws. These were replaced, some generations ago, by beings who once, in another life, were ladies; once were Europeans. To these lady-like women in the wilderness there is something precious and unnatural about lady-likeness, about the cultivated European rose grafted upon the wild American stem. But "pretense had always been their aim," even when in childhood they had played house, played grown-up. Their conscious female end is that genteel, cultured, feminist old-maidishness that—intact, thorny, precisely self-contained—rises above the masculine, disreputable economic and sexual necessity that reaches out to strip off their blossoms, that makes you "dish potatoes up three times a day,/ And put your wedding dress into a quilt," that turns young ladies into old women. This Victorian old-maidish culture has its continuation in the *House Beautiful*, *Vogue*-ish sophistication that the poet calls "our exotic properties, our pretty price./ The garden radish lies on ice, the radish rose./ Smorgasbord!" The new *déjeuner sur l'herbe* is summed up, bitterly, in the old terms, the plain, religious, country terms:

"Dinner on the grounds! and the blessing still unsaid . . ." The poet looks askance at this acquired surface, even in herself—especially in herself, since it belongs neither to her wild heart nor to her neo-Calvinist conscience. To her there is something natural and endearing about the crushing wilderness, the homely childish beauties that one relaxes or regresses into. It is the ladies who really are barren, so that one might say: "You make a desert and call her a lady"; and in what is perhaps the most beautiful and touching of all these poems, *Buck Duke and Mamma*, it is human feeling, natural sexuality, that the woman at last accepts in grief, and it is the histrionic feminine gentility that she rejects:

He came bringing us a milkpail full
Of speckled, wild, goose plums—
All fat unsmelt-out perfumedom—
And perched on the back porch curb to taste a few.
"Sour! Your eyes'll water, Miss Tempe!
But sweet, too."
Mamma's way was posing by the silent pool
And tossing in the line amiss
That shook the skies of the other world
And all but loosed the roots of this.
She trimmed and trained the roundabout backwoods,
Was glad that Buck Duke had a devilish eye;
It saved an orphan from dire fortitude,
And saved his grandpa's house from sanctity.
"Your Papa doesn't favor your going there.
I say, enter evil to cure evil, if you dare!"
As she went about her cast-off household chores
She overlooked them with a lavish bow
Inspired by that heroine of poems,

Her elocution teacher, Miss Hattie Yow.
"Nothing to do? In this world of ours?
Where weeds spring daily amidst sweet flowers?"
Your-mammy-never-came-to-much-my-Buck.

"Don't drink that Mackling Spring's brack water
Whe'r it's high or low.
The cows stand in there and let go."
But old Duke's beardy words were moss for campfire
When they took their kitchen rations to the woods.
Mamma's boys looked out for sassafras, but Buck
Made frog gigs, thrashed Mackling Springs into a suds.
("I say, dear boys! Be good. Take care.
But learn a little evil if you dare. . . .")
His thirst once drunk, turned drunken,
And Buck Duke tossed all night, all day,
Made rusty speeches on old swapping knives,
Called names that paled the sallow-boned herbwives,
Tore off the sleeping clothes, his bed's, his own,
And never seemed to wake.
His boyish modesty ran dry,
At last the hands cooled, then the face.
Mamma stood at his bedside.
She overlooked him with a sprightly brow
Inspired by that gay mistress of mad poesy
Her elocution teacher, Miss Hattie Yow.
" 'Stop stop pretty waters' cried Mary one day
'My vessel, my flowers you carry away.' "

Mamma made a wreath of all her flowers:
The histrionic garden did not bear
One saucy pose when she put down the scissors;
The battered bees hung stupid in mid-air.
She worked on knees and elbows on the back porch,
That savage zinnia ornament compiled,
Then all at once cooped up her face

With hands like bird's wings—
A gesture, she knew, would have made Miss Hattie
smile.

If these poems are less about the New Woman than about the Old (surviving, astonished, into this age of appliances and gracious fun), still, no poems can tell you better what it is like to be a woman; none come more naturally out of a woman's ordinary existence, take both their subjects and their images out of the daily and nightly texture of her life. Many of these are what I think of as woman's-work-is-never-done images: cooking, sewing, ironing, taking care of children, tending the sick, and so on; but these pass, by way of gardening, on over into the lady-like images of social existence, distrusted things akin to all the images of glass, mirrors, gems, of coldness, hardness, and dryness, of two-ness, cleavages, opposites, negatives, of trapped circular motion, that express a range of being from gentility to catatonia. These are lightened, colored, by images from childhood and the past—counting-out rhymes, hymns, slave songs, and so on. A pervading, obsessive image is that of light in darkness: there are so many stars, meteors, flames, snowflakes, feathers, that one almost feels that the poems themselves can be summed up in the sentence in which the dying old woman sums up her life: "My quick, half-lighted shower, are you gone?"

Often these last images merge with the ruling, final image of the poems, that of water: the water of experience or sexuality, into which the little girl is afraid to wade; the river the dying woman remembers from childhood and must cross, now, into the next world; life's dark star-bearing flood trapped in the mill of daily duties, of reduced mechanical existence: "The water pushes the mill wheel;/ The wheel, wheeling, dispersing,/ Disperses the starry spectacle/ And drags the stone" —trapped, or else frozen into the fixity of glass, of mirrors, of the hateful gems that send the hands on in their aimless endless circle. Even sexual love is seen in terms of water freezing into—or melting among—the "thin floes," the cold clandestine darkness of a country night. You destroy yourself, escape from yourself, in water: in *Goodbye Family*, "under the foundations of God's World/ Lilily/ Swimming on my side" until at last "the water/ Meeting me around the curve, roaring, blanks/ Out all"; and in *Escape*, as "a vein of time gapes for her small transfusion," she or her double disappears into the ocean with a "far white crash too negligible to bear." She says that "art and death" are "both oceans on my map"—the map of the *Woman as Artist*, woman as lover. And woman ends, man ends, "lying at the edge of the water," face in the water; "when our faces are swol up/ We will look strange to them./ Nobody, looking out the door/ Will think to call us in./ They'll snap their fingers trying/ To recollect our names"; the rope is broken, no one ever again will draw up the bucket bobbing at the bottom of the well of death:

Oh my dearie,
Our childhoods are histories,
Buckets at the bottom of the well,
And hard to tell
Whether they will hold water or no.
Did Pa die before we were married?
No, he died in twenty-seven,
But I remember the wedding
Reminded me of the funeral—
When the grandbabies ask,
Little do they care,
I will tell them about the man I found
That day at my plowing in the low-grounds
Lying at the edge of the water.
His face had bathed five nights.
A dark man, a foreigner, like.
They never found his kin to tell.
Buckets, buckets at the bottom of the well.
It was in the paper with my name.
I have the clipping tells all about it,
If your Grandma aint thrown it out.

Oh my dearie
When our faces are swol up
We will look strange to them.
Nobody, looking out the door
Will think to call us in.
They'll snap their fingers trying
To recollect our names.
Five nights, five bones, five buckets—
Who'll ever hear a sound?
Oh my dearie
The rope broke
The bucket bobs around
Oh my dearie

In the poems everything goes together, everything has several reasons for being what it is: the whole *Wilderness of Ladies* is, so to speak, one dream, that expresses with extraordinary fidelity and finality the life of the dreamer.

Many of the poems show (rather as the end of *The Old Wives' Tale* shows) what you might call human entropy—life's residual reality, what is so whatever else is so. That life, just lived, is death; that its first pure rapturous flame grows greater, fouls itself, diminishes, struggles and goes out: the poems say it with terrible magic:

In the morning, early,
Birds flew over the stable,
The morning glories ringed the flapping corn
With Saturn faces for the surly light
And stars hung on the elder night.

But soon the sun is gone, the stars go out as the old woman's eyes close. Life is a short process soon over: how quickly the lyric, girlish, old-fashioned funniness of *My Grandmother's Virginhood, 1870* becomes the worn, sad grown-upness of *Motherhood, 1880!*—the girl's kiss so soon is the woman's sick or dying baby, her eroded featureless "They know I favor this least child."

The poems are full of personal force, personal truth—the first and last thing a reader sees in a writer—down to the least piece of wording. Their originality is so entire, yet so entirely natural, that it seems something their writer deserves no credit for: she could do no other. Just as the poems' content ranges from pure fact to pure imagination, so their language ranges from a folk speech as authentically delightful as Hardy's or Faulkner's (though the poet's use of folk material reminds me even more of Janacek's and Bartok's) to a poetic style so individual that you ask in wonder: How can anything be so queer and yet so matter-of-fact—natural, really? Picasso has said that when you find the thing yourself it is always ugly, the others after you can make it beautiful. Sometimes this is true of these poems; and yet sometimes she has found it beautiful, or has made it into a marvel you don't call either beautiful or ugly—have no words for:

Was it forgiven? It was gone,
The heathen dancing
With her giggling sisters;
They flew about the room
In seedstitch weskits
Like eight wax dolls gone flaskwards.
Those were gay days!
She sighed a mournful tune
Waddling about her everyday
Affairs of life and death
(*Affairs of painful life, uncertain death*):
"*Wild loneliness that beats*
Its wings on life," *she sang.*
She thwacked a pone in two,
Her big hand for a knife.
Thar! stirring it severely,
And thar! into the oven. . . .

There is plenty of detached objective observation in the poems, but usually they are objective in another sense: they are so much the direct expression of the object that their words are still shaking with it—are, so to speak, *res gestae*, words that, repeated, are not hearsay evidence but part of the fact itself. The poet continually makes a kind of inevitable exclamation, has wrenched from her a law or aphorism, a summing-up, that is at the same time an animal cry. Sometimes her speech is the last speech before speechless desperation—too low to be heard as sound, only felt as pain; but sometimes it is like sunlight on fall leaves, firelight on cornbread. The book presents as they have never been presented before—which is to say, as every true artist has presented them—our everyday affairs of life and death.

Some of the very best of Eleanor Taylor's poems, I think, are *Buck Duke and Mamma*, *Song*, *Woman as Artist*, *Moved*, *Family Bible* (especially *Grandparents*), *The Bine Yadkin Rose*, and *Goodbye Family;* poems like *Madame*, *In the Churchyard*, *The Chain Gang Guard*, and *Playing* are slighter or smaller, but realized past change. Readers who are well acquainted with all of *Wilderness of Ladies* will feel an impatient disgust at me for some of the poems I haven't named, the qualities I haven't mentioned. And all the poems are far more than the best poems: the pieces, put together, are a world.

When one reads poems here and there, in magazines and manuscript—as I first read these—it seems very unlikely that they should be good almost as Dickinson's or Hardy's poems are. Of course the readers who first saw Dickinson's and Hardy's poems, in magazines and manuscripts, thought it just as unlikely that the poems should be good almost as Wordsworth's were. The readers knew what the poems weren't, what the poems couldn't be; and because of this it was hard for them to see what the poems were. An introduction to poems like those in *Wilderness of Ladies* might make it easier for readers to consider the possibility of the poems' being what they are.

Randall Jarrell

was born in Nashville, Tennessee in 1914 and graduated from Vanderbilt University. He now lives with his wife and two daughters in Greensboro, North Carolina, where he is Professor of English at the Woman's College of the University of North Carolina. Mr. Jarrell has also taught at Sarah Lawrence and Kenyon Colleges, the Universities of Texas, Illinois, Indiana and Cincinnati, and at Princeton University. At various times he has been poetry critic of the Nation, Partisan Review and The Yale Review, and as poet, novelist and critic his work has received many awards. For two years he was Consultant in Poetry at the Library of Congress. He is a member of the National Institute of Arts and Letters and a chancellor of The Academy of American Poets. His books include six volumes of poems, among which is the *Selected Poems* (1955) and *The Woman at the Washington Zoo* (which received the 1961 National Book Award for Poetry), a work of fiction (*Pictures from an Institution*), and an earlier book of essays, *Poetry and the Age.*

PS3519 A86 S3
+A sad heart at t+Jarrell, Randall
0 00 02 0204390 7
MIDDLEBURY COLLEGE